# Lecture Notes in Mathematics

Edited by A. Dold and B. Eckmann

1282



David E. Handelman




# Positive Polynomials, Convex Integral Polytopes, and a Random Walk Problem


Springer-Verlag

Berlin Heidelberg New York London Paris Tokyo

**Author**

David E. Handelman
Mathematics Department, University of Ottawa
Ottawa, Ontario K1N 6N5, Canada

Mathematics Subject Classification (1980): 06F25, 13B99, 19A99, 19K14, 46L99, 52A43, 60G50

ISBN 3-540-18400-7 Springer-Verlag Berlin Heidelberg New York
ISBN 0-387-18400-7 Springer-Verlag New York Berlin Heidelberg


Printed in Germany

Printing and binding: Druckhaus Beltz, Hemsbach/Bergstr.
2146/3140-543210

# POSITIVE POLYNOMIALS, CONVEX INTEGRAL POLYTOPES, AND A RANDOM WALK PROBLEM

## PROSPECTUS

This monograph concerns itself with results and interconnections in a number of areas; these include positive polynomials, a class of special random walk problems on the lattice $\mathbf{Z}^d$ (d an integer), convex integral polytopes (that is, convex polytopes in $\mathbf{R}^d$ all of whose vertices are lattice points), reflection groups, and commutative algebra. Techniques include those of functional analysis (especially Choquet theory), ordered rings, commutative algebra, and convex analysis. The central problems arise from special actions of tori on C*-algebras, and many of the results in the other areas yield results back at the C* level; the translation is implemented by means of ordered $K_0$ (of the fixed point C*-algebras). These connections are (roughly) described in Table 1. Since the techniques and results deal with a number of disparate fields, we shall develop the material in each area in considerable detail, as not everyone will be familiar with all of the topics discussed.

This prospectus is intended to outline the various interconnections; topics which may not be familiar to the reader will be introduced in the body of the text. In this prospectus I am trying to make the case that there is a lot of interesting mathematics occurring here, and that the scope for further research is vast.

The motivating problem arises from the classification and description of invariants arising from "xerox" type actions of tori on C*-algebras. Specifically, let $\pi: \mathbf{T} \longrightarrow U(n,C)$ be an n-dimensional representation of the d-torus $\mathbf{T}$; let $A = \otimes M_n C$ (n fixed) be the infinite tensor product of $n \times n$ matrix algebras, and define $\alpha: \mathbf{T} \longrightarrow \mathrm{Aut}(A)$ via $\alpha(g) = \otimes \mathrm{Ad}\pi(g)$. Then the fixed point algebra of A under the action of $\alpha$, $A^{\mathbf{T}}$, is of great interest. The classification of such objects is done by means of ordered $K_0(A^{\mathbf{T}})$. Consequences of this include the parameterization of all primitive ideals of the fixed point algebras, including explicit generating sets for these ideals. Moreover, the algebraic structures developed herein yield invariants (essentially complete and computable) for $A^{\mathbf{T}}$.

In order to compute the ordering on $K_0(A^{\mathbf{T}})$ (which is the crucial ingredient of this invariant), one has to solve problems of the following type. Given (Laurent) polynomials P and f in several (real) variables, with the coefficients of P being non-negative, determine necessary and sufficient conditions on f for there to exist an integer n so that $P^n f$ also has no negative coefficients. (Reference [H1] is primarily devoted to this, and the solution is given in [H2].) Here P corresponds to the character of $\pi$.

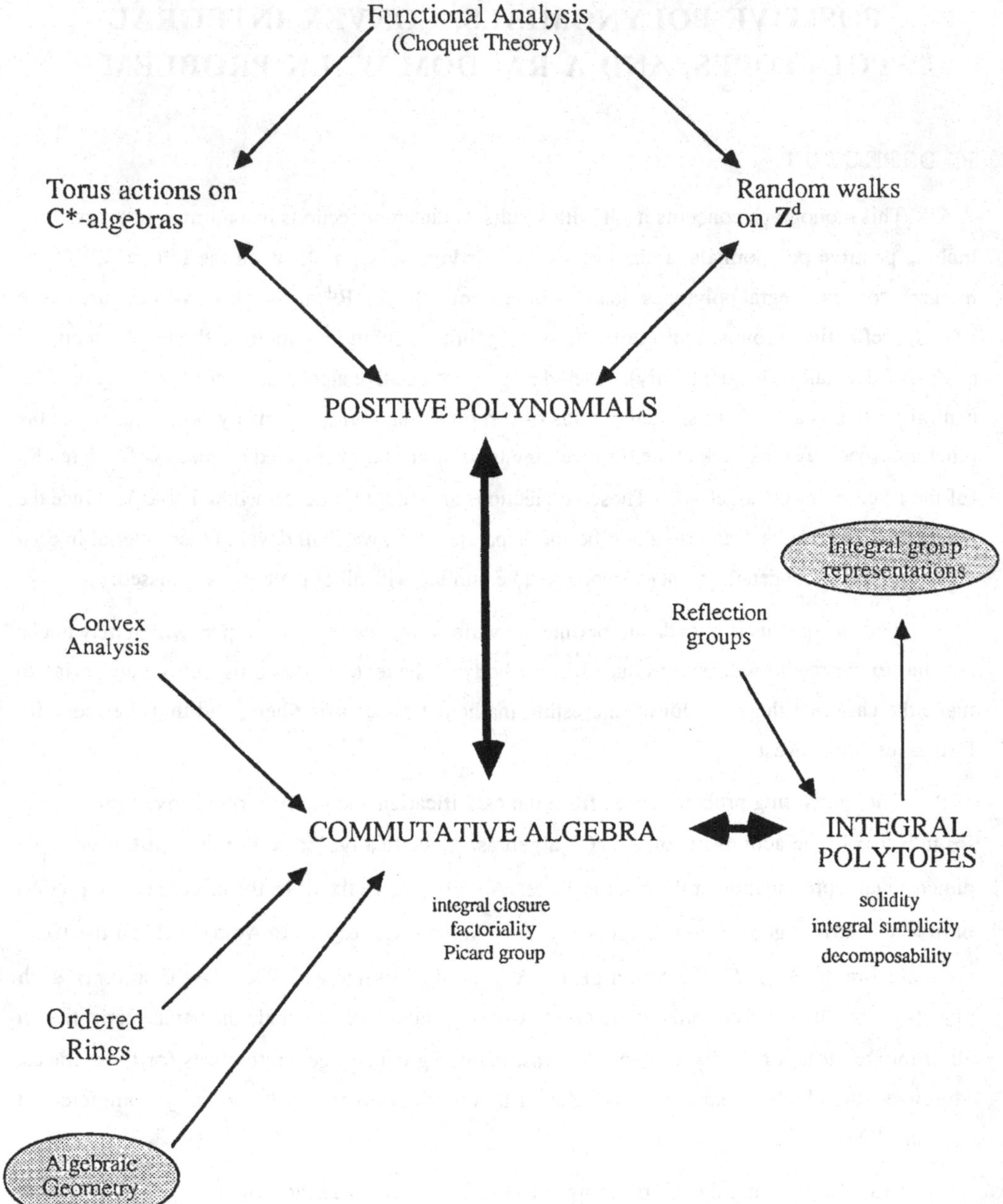
Functional Analysis
(Choquet Theory)
Torus actions on
C*-algebras
Random walks
on Z^d
POSITIVE POLYNOMIALS
Integral group
representations
Convex
Analysis
Reflection
groups
COMMUTATIVE ALGEBRA
INTEGRAL
POLYTOPES
integral closure
factoriality
Picard group
solidity
integral simplicity
decomposability
Ordered
Rings
Algebraic
Geometry

Table 1

For example, consider what happens when $d = 1$, that is, $P$ and $f$ are polynomials in a single variable $x$. Provided $P$ is not trivial and has no gaps (this means that if $x^a$ and $x^c$ appear in $P$, and $a < b < c$ with $b$ an integer, then $x^b$ appears in P), the conditions on $f$ are otherwise independent of the choice of P; $f$ is strictly positive as a function on the interval $(0,\infty)$. (This can be deduced from a result of Meissner [M], and is also proved in [H1]). Similarly, whether $P = x^3 + x + 1$ or $P = x^3 + 2x + 4$ (or any $P = x^3 + dx + e$, with $d$ and $e$ greater than 0), the conditions on $f$ depend only on the set of supporting monomials in P. In this case, the conditions are that $f$ be strictly positive on the interval $(0,\infty)$ and that the coefficient of $x^{\deg f - 1}$ be non-negative [H2; Example V.2].

In several variables, the conditions on $f$ (given P) are so complicated [H2], that it is not clear that this principal applies; in other words, if $P_0$ and $P_1$ are polynomials with only positive coefficients having identical sets of supporting monomials, is it true that $P_0^n f$ has only positive coefficients (for some n) if and only if there exists $m$ so that $P_1^m f$ has as well?

If we translate this (via the inverse Fourier transform) to a question about random walks on $\mathbf{Z}^d$, then this would seem to be counter-intuitive, Let $\nu_f$ be a signed measure of finite support (the correspondence between $f$ and $\nu_f$ being given by $f = \sum \nu_f(w) x^w$) and let $\mu_o$ and $\mu_1$ be finitely supported positive measures on $\mathbf{Z}^d$. Is it true that with $*$ denoting convolution, $(\mu_o * \mu_o * \ldots * \mu_o) * \nu_f$ being positive for some number of $\mu_o$'s implies that $(\mu_1 * \mu_1 * \ldots * \mu_1) * \nu_f$ is positive as well (for a sufficiently large number of $\mu_1$'s)? For example, suppose that the support of $\mu_o$ and of $\mu_1$ is the set of vertices of the unit square in $\mathbf{Z}^2$, and most of the mass of $\mu_o$ is concentrated at (0,0), while most of the mass of $\mu_1$ is at (1,1). It seems unreasonable that the answer to our question should be affirmative, because each measure is distorted in a different direction. Nonetheless, the result *is* true, and is even true with a slight weakening of the hypotheses. This is the principal result of section II, and forms much of the basis for the rest of the paper. The techniques involve a variant of Choquet theory developed in [GH1]) for dimension groups, as well as those of commutative algebra.

This result permits us to define an AGL(d,Z)-invariant for integral polytopes, as follows.

Let $K$ be an integral polytope in $\mathbf{R}^d$. Let $P$ be any polynomial with only positive coefficients, whose set of supporting monomials is $K \cap \mathbf{Z}^d$; that is, write $P = \sum \lambda_w x^w$ (here $w = (w(1), w(2), \ldots, w(d))$ is a point in $\mathbf{Z}^d$ and $x^w = x_1^{w(1)} x_2^{w(2)} \ldots x_d^{w(d)}$), with all of the $\lambda_w$ being non-negative real numbers such that $\lambda_w$ is not zero if and only if $w$ belongs to $K \cap \mathbf{Z}^d$. Form the ring $R_P = \mathbf{R}[x^w/P;\, w \in K \cap \mathbf{Z}^d]$ (this is a subalgebra of $\mathbf{R}[x_i^{\pm 1}, P^{-1}]$) with positive cone generated additively and multiplicatively by $\{x^w/P \mid w \in K \cap \mathbf{Z}^d\}$. If $f$ is a polynomial with $f/P^k$ in $R_P$ for some integer $k$, then $f/P^k$ belongs to the positive cone if and only if there exists $n$ so that $P^n f$ has no negative coefficients. Let U denote the set of elements $\{a = f/P^k \in R_P | a(r_1,...,r_d) > \varepsilon$ for some $\varepsilon$, but for all strictly positive real $r_1, ..., r_d\}$. This is precisely that set of elements of $R_P$ that are bounded below by a positive rational multiple of the constant function 1 with respect to the ordering. Form $U^{-1}R_P$ (i.e., invert all of the members of the multiplicatively closed set U). The preceding random walk result yields that $U^{-1}R_P$ depends *only on* $K$ (as an ordered ring) and not upon the choice of P. Set $R_K = U^{-1}R_P$; this is easily seen to be an AGL(d,Z)-invariant for K. In fact we treat somewhat more general objects arising from arbitrary finite subsets of $\mathbf{Z}^d$, instead of $K \cap \mathbf{Z}^d$.

The obvious thing to do, is to develop a lexicon between (algebraic) properties of $R_K$ and integro-geometric properties of K. We deal with several properties of $R_K$, e.g., being integrally closed, Cohen-Macauley, or factorial (and for $R_P$, being regular). In each case, interesting subsidiary results come out. For example, a standard easy to prove fact is that if $\{a_i\}$ is a finite set of positive integers whose greatest common divisor is 1, then all sufficiently large integers are positive integer combinations of the $a_i$. The higher dimensional analogue is neither obvious nor (apparently) known. If A is a subsemigroup of $\mathbf{Z}^d$ (under addition) such that $A-A = \mathbf{Z}^d$, and b is a lattice point in the interior of the convex hull of A, then there exists an integer m so that for all $n \geq m$, nb belongs to A. Strange phenomena occur along the boundary of the convex hull of A, and therein lies the obstruction to $R_{dK}$ being integral over $R_K$ in general ($d = \dim K$).

One rather surprising phenomenon occurring only when the dimension is at least three, is that $2(K \cap \mathbf{Z}^d) \neq 2K \cap \mathbf{Z}^d$, i.e., there exists w in $2K \cap \mathbf{Z}^d$ that is not a sum of two (or frequently, an arbitrary number of) elements of $K \cap \mathbf{Z}^d$, even when $K \cap \mathbf{Z}^d - K \cap \mathbf{Z}^d$ generates $\mathbf{Z}^d$ as an abelian group. An integral polytope is called solid when $n(K \cap \mathbf{Z}^d) = nK \cap \mathbf{Z}^d$ for all n, and a local version of this is precisely the criterion (when $K \cap \mathbf{Z}^d - K \cap \mathbf{Z}^d$ generates $\mathbf{Z}^d$) for $R_K$ (and $R_P$) to be integrally closed (and Cohen-Macauleyness also occurs). It turns out that solidity is frequently easy to verify, as $n(K \cap \mathbf{Z}^d) = nK \cap \mathbf{Z}^d$ is sufficient, as is $K = eK'$ for some $e \geq d$ and K' an arbitrary integral polytope.

One of the principal tools is the Riesz interpolation property; this is satisfied by $R_P$ (when viewed as an ordered abelian group under addition). This allows us to show that every localization of $R_P$ at a prime ideal is also a localization of a monomial algebra, hence is amenable to results of Hochster [Ho].

A first step in determining conditions for factoriality of $R_K$ (and regularity of $R_P$), and which is of interest in its own right, is to show that all projectives over $R_K$ (arbitrary K) are free. To do this, we use the Legendre transformation from convex analysis. Let $P = \sum \lambda_w x^w$ be a polynomial in d variables with no negative coefficients. Define the map $\Lambda: (\mathbf{R}^d)^{++} \longrightarrow \mathrm{Int}\, K$ via

$$\Lambda(r) = \frac{\sum \lambda_w r^w w}{\sum \lambda_w r^w} = \left( \ldots, \frac{x_i \frac{\partial P}{\partial x_i}}{P} \Big|_{x=r}, \ldots \right)$$

(Here, $r^w = r_1^{w(1)} r_2^{w(2)} \ldots r_d^{w(d)}$.) If K has interior, then a very special case of [Ro; Theorem 26.5] yields that $\Lambda$ is a homeomorphism. This beautiful result, also known to probabilists [N], [INN] is apparently unknown to differential and algebraic topologists (as an informal survey of more than 20 revealed!). So I have included a proof of this result using only elementary techniques from real and complex analysis, functional analysis, and operators on Hilbert space (Appendix E).

The connections between the Legendre transformation and projectives over $R_K$ run as follows. If G is a partially ordered abelian group, a state of G is a positive real-valued group homomorphism. The states of $R_K$ normalized at 1, form a Choquet (in fact, a Bauer) simplex. Among them are point evaluations arising from r in $(\mathbf{R}^d)^{++}$, given by $a \mapsto a(r)$ for a in $R_K$. It is known that the set of these is dense in the set of pure (= extremal) states, which we denote $d_eS(R_K,1)$. For any choice of P that will generate K, there is a natural map, $\Lambda_P: d_eS(R_K,1) \longrightarrow K$ which when restricted to $(\mathbf{R}^d)^{++}$ (viewed as the set of point evaluation states) is simply $\Lambda$. It is known that $d_eS(R_K,1) \setminus (\mathbf{R}^d)^{++}$ is a union of lower dimensional spaces that are each state spaces of a similar type (arising from polynomials in effectively fewer variables than P), and from $\Lambda$ being a homeomorphism, it follows inductively that $\Lambda_P$ is as well. Thus $d_eS(R_K,1)$ is homeomorphic to K, so in particular, is contractible.

Now $R_K$ embeds naturally as a ring of continuous (real-valued) functions on $d_eS(R_K,1)$ (this is the standard method of analyzing dimension groups—by representing them as groups of affine functions on the state space), and moreover, everything in $R_K$ that does not admit a zero (as a function) on the compact space $d_eS(R_K,1)$ has an inverse in $R_K$. Standard arguments (as in [Sw1]) then yield that all projectives are free, as a consequence of contractibility of $d_eS(R_K,1)$.

Factoriality of $R_K$ reduces to regularity of $R_P$. This turns out to be equivalent (when $K \cap \mathbf{Z}^d - K \cap \mathbf{Z}^d$ generates $\mathbf{Z}^d$ as an abelian group) to the following very strong geometric property for K:

> For every vertex of K, the convex hull of that vertex together with its nearest neighbours in $K \cap \mathbf{Z}^d$ along the edges (1-faces) of K that contain the vertex, is AGL(d,Z)-equivalent to the standard solid d-simplex.

There is a very simple test for an integral polytope to be AGL(d,Z)-equivalent to the standard simplex—it is simply that the volume be 1/d!. An integral polytope K satisfying the property above will be called integrally simple; this notion is the integral analogue of simplicity for real polytopes. We study integrally simple polytopes in some detail. Their key property is that they admit "local integral coordinates". This may be exploited, for example, to determine all the meet-irreducible order ideals of $R_K$; this translates back to the determination of all primitive ideals in the original fixed point C* algebra $A^T$ that we discussed in the beginning.

Such a strong property might seem to render this class too restrictive. However, integrally simple polytopes occur in profusion, as the examples in Appendix D demonstrate. Let W be the Weyl group of some compact connected semisimple Lie group. It acts in a natural way on the dual of the maximal torus, i.e., on $\mathbf{Z}^d$. Let a be any lattice point, and let $K_a$ be the convex hull of the W-orbit of a. Regarded (after translation) as inside the sublattice of $\mathbf{Z}^d$ generated by $\{a^g - a \mid g \in W\}$, $K_a$ is integrally simple (with respect to this sublattice) for almost all choices of a; in particular, $K_a$ is integrally simple if a is not on the boundary of the Weyl chamber to which it belongs. This type of result also holds, with restrictions, for some reflection groups.

This result suggests a geometric invariant for general finite group integral representations—given a group acting on $\mathbf{Z}^d$, look at the convex hulls of the orbits, and ask questions about their AGL(d,Z)-invariant properties, either individually, or as above en masse. We do not investigate this here (which explains the shaded oval in table 1).

Automorphisms of $R_K$ are of interest. We show that if K is either integrally simple or indecomposable (in the sense of real polytopes, that is, K = K' + K" implies one of K' or K" is real homothetic to K), then any algebra automorphism of $R_K$ is automatically an order-automorphism (section VIII). While this would seem to indicate that the order structure can be done away with (and there is no reason to believe the result fails for more general K), in fact, we use order-theoretic methods and concepts throughout, and these provide us with indispensable tools and points of view. It follows from this result, that for these K, the automorphisms of $R_K$ are of a very special form. Aside from a finite group of symmetries on K, there is a collection of automorphisms associated bijectively with the faithful pure states of $R_K$, i.e., with $(\mathbf{R}^d)^{++}$. These (dropping the finite top) appear to be the only ones, and this conjecture leads to what seems to be a difficult problem in algebraic geometry, concerning special automorphisms of function fields. Except in the relatively trivial case of d = 1, I could not solve this, explaining the encircling of algebraic geometry in the table.

In connection with indecomposables (or rather decomposables), there is a rather startling relation between decompositions, factorization of polynomials in several variables, and the Picard group of $R_P$. Let P (with only positive coefficients) be such that its set of exponents arising from supporting monomials is of the form $K \cap \mathbf{Z}^d$, where K is integrally simple. Then $R_P$ is (homologically) regular, and its Picard group, $\mathrm{Pic}(R_P)$ is generated by prime (order) ideals corresponding to faces of codimension 1 in K.

Essentially, the obstructions to $\mathrm{Pic}(R_P)$ being trivial are given by a failure of P to admit a factorization corresponding to a decomposition of K as a sum of other (possibly homothetic) rational polytopes (which need not be simple themselves). In particular, $\mathrm{Pic}(R_P)$ is as large as possible (with K fixed) when P is irreducible (and there is a simple formula in terms of d and the number of maximal faces), and gets smaller as P starts having factorizations. A careful examination of how factorizations of P lead to new relations among the generators of $\mathrm{Pic}(R_P)$ yields that integrally simple polytopes decompose into the known maximal number of indecomposable pairwise non-homothetic rational polytopes (i.e., the theoretical upper bound is attained). (This may be known even in the more general context of simple polytopes, but I could not find any such result in the literature.) The argument indirectly requires the Legendre transformation, of course!

Frequently it happens that $\mathrm{Pic}(R_P) = 0$, which upon translation back to a corresponding fixed point C* algebra $A^T$ yields the bizarre

$$K_0(K_0(A^T)) = \mathbf{Z}$$

for this particular fixed point algebra.

Scattered throughout the monograph are elaborations on the theme of the random walk problem. For example, if f is in $\mathbf{R}[x_i^{\pm 1}]$ and there exists Q in $\mathbf{R}[x_i^{\pm 1}]^+$ having no negative coefficients such that the same holds for $P = Qf$, does it follow that there exists n so that $P^n f$ has no negative coefficients as well? Easy examples provide a negative answer. However, if $R_P$ is integrally closed (and necessary and sufficient geometric conditions are determined in section III for this to occur) the answer is affirmative.

Suppose P and Q are in $\mathbf{R}[x_i^{\pm 1}]^+$ and Q is obtained from P by deleting a codimension 1 face from the K corresponding to P (that is, $K = \mathrm{cvx\ Log}\ P$). If $Q^n f$ has no negative coefficients for some n (with f in $\mathbf{R}[x_i^{\pm 1}]^+$), does it follow that the same holds for $P^m f$ for some m? Again there are plenty of counter-examples. However, if $R_P$ is homologically regular (which translates more or less to its corresponding K being integrally simple), the answer is yes!

Now suppose that P is a fixed polynomial (in several variables) with only positive coefficients, and let $\{P_1, P_2, ...\}$ be a sequence of such so that (i) the monomials appearing in each $P_i$ (with nonzero coefficient) are exactly the same as those of P, and (ii) the set of *all* nonzero coefficients of all the $P_i$'s is bounded above and below (away from zero). For f in $\mathbf{R}[x_i^{\pm 1}]^+$, if $P^n f$ has no negative coefficients for some n, there exists m so that $P_1 P_2 \ldots P_m f$ has the same property. The obvious convergence argument just fails to prove this, but a trick involving the thematic random walk result allows it to be pushed through (the converse is a trivial consequence of the random walk result).

We return occasionally to the motivating fixed point C*-algebras. As was mentioned earlier on, when the corresponding $K$ is integrally simple (e.g., if the representation $\pi$ of $T$ is the restriction to the maximal torus of an irreducible character of a compact connected semisimple Lie group with dominant weight not on the boundary of the Weyl chamber), the primitive ideals, their structure, subquotients, and generators, can be determined merely by translating back the results on the meet-irreducible order ideals of $R_K$ (section VII). For those who are still interested, we also give a C* algebraic proof of a result in [H1] dealing with the quotients of $A^T$ corresponding to faces of K (Appendix B).

# TABLE OF CONTENTS

## INTRODUCTION

Let $K$ be a compact convex polytope in $\mathbf{R}^d$, all of whose vertices have only integer coordinates; we say $K$ is an integral polytope. Let $AGL(d,\mathbf{Z})$ be the group of transformations of $\mathbf{Z}^d$ generated by $GL(d,\mathbf{Z})$ and translations by elements of $\mathbf{Z}^d$. We construct a partially ordered commutative ring $R_K$ which is an $AGL(d,\mathbf{Z})$ invariant (that is, if $K$ and $K'$ are two such polytopes and there is $g$ in $AGL(d,\mathbf{Z})$ with $gK = K'$, then $g$ induces an isomorphism of ordered rings $R_K \longrightarrow R_{K'}$). The construction emanates from the classification of fixed point C*-algebras, $A^T$, under what have become known as xerox type actions of d-tori (see for example, [HR]).

This classification has been given by means of $K_0(A^T)$; this is a finitely generated subring of the function field $\mathbf{Q}(x_i)$ in d variables. In what follows, we actually work with real rather than rational coefficients, but virtually all results that hold with real coefficients hold with rational ones and vice versa. Corresponding to the action of the d-torus, is a character thereof, $P$, in $\mathbf{Z}[x_i^{\pm 1}]^+$, the plus sign indicating that all of the coefficients of $P$ are non-negative. Then $K_0(A^T)$ is

$$R_{P,\mathbf{Z}} = \mathbf{Z}[x^w/P \mid w \in \mathbf{Z}^d,\ x^w \text{ appears in } P],$$

with positive cone generated additively and multiplicatively by

$$\{x^w/P \mid w \in \mathbf{Z}^d,\ x^w \text{ appears in } P\} \qquad [H1;\ I.4].$$

Here we are using the notation $x^w = x_1^{w(1)} x_2^{w(2)} \dots x_d^{w(d)}$ where $w = (w(1),\dots,w(d))$ belongs to $\mathbf{Z}^d$. If $\mathbf{Z}$ is replaced by $\mathbf{R}$, we define, for $P$ in $\mathbf{R}[x_i^{\pm 1}]^+$,

$$R_P = \mathbf{R}[x^w/P \mid w \in \mathbf{Z}^d,\ x^w \text{ appears in } P],$$

with positive cone $R_P^+$, generated additively and multiplicatively over $\mathbf{R}^+$ by $\{x^w/P \mid w \in \mathbf{Z}^d, x^w$ appears in $P\}$.

To obtain our invariant, the partially ordered ring $R_K$, we initially set $P = \sum x^w$, the sum being taken over $K \cap \mathbf{Z}^d$. Define $R_P$. Then we invert all the elements of $R_P$ that are order units, that is, positive elements bounded below by a positive rational multiple of 1, either as functions on $(\mathbf{R}^d)^{++} = \{r = (r_1,\dots,r_d) \in \mathbf{R}^d \mid r_i > 0\}$, or with respect to the ordering described above (the two are equivalent). These elements have been described in reasonably detailed fashion [H1; section V]. This definition already leads to a problem, concerning the ordering on $R_K$.

There is a natural direct limit ordering that can be put on $R_K$, and it is not at all clear that if $a$ in $R_P$ becomes positive in $R_K$ (in other words, if $au$ is positive in $R_P$ for some order unit $u$), then $a$ is already positive in $R_P$. In fact this is the case; moreover, we can replace our original choice of $P$ used to define $R_K$ by any $P$ in $\mathbf{R}[x_i^{\pm 1}]^+$ such that:

$$\{w \in \mathbf{Z}^d \mid x^w \text{ appears in } P\} = K \cap \mathbf{Z}^d.$$

The set on the left (defined for any Laurent polynomial, not necessarily with positive coefficients) is called Log P.

This positivity problem is really a random walk problem on the lattice $\mathbf{Z}^d$, as follows. Let f be an initial finite distribution of real masses (allowing negative ones) on $\mathbf{Z}^d$, and let P and Q be finite distributions of positive masses. Let $P^n f$ denote the distribution obtained by convolving P with itself n times and then once with f. If the support of P equals that of Q, then a special case of Theorem II.1 (which is enough to resolve the ambiguous positivity problem described above) asserts that there exists n so that $P^n f$ is a positive distribution if and only if there exists m so that $Q^m f$ is also positive. By considering P, Q and f as the Laurent polynomials they represent (via the inverse Fourier transform), we obtain a translation to an eventually positive result (here for example, positive distribution translates to positive coefficients). On the other hand, if the support of P differs from that of Q, for example, if Log P is the set of vertices of the unit square in the plane and Log Q is the set of vertices of the unit triangle, then there is not even a one-sided implication about eventual positivity.

The third chapter discusses the integral closure of $R_K$ (and $R_P$) in its field of fractions, and related notions. Call an integral polytope "solid" if for all integers k, $k(K \cap \mathbf{Z}^d) = kK \cap \mathbf{Z}^d$. If d = 2, then all integral polygons are solid, but in higher dimensions even under a faithfulness assumption (that $K \cap \mathbf{Z}^d - K \cap \mathbf{Z}^d$ generate all of $\mathbf{Z}^d$ as an abelian group), this fails. It turns out that dK (d being the dimension of the Euclidean space in which K is embedded) is always solid, and moreover, $dK \cap \mathbf{Z}^d = d(K \cap \mathbf{Z}^d)$ is sufficient for K to be solid. If K is solid, then $R_K$ is integrally closed and Cohen-Macauley (the latter possibility was suggested by M. Hochster). There is actually an ostensibly stronger result, using a notion of local solidity.

We also determine when $R_K$ is an order in $R_{dK}$; this is equivalent to the integral closure of $R_K$ being of the form $R_{K'}$, for some integral polytope K'. If d = 3, this is automatic, but counter-examples exist in all higher dimensions. The criterion is a sort of local faithfulness condition, specifically, that for all faces F of K, the group generated by $dF \cap \mathbf{Z}^d - dF \cap \mathbf{Z}^d$ equal that generated by $F \cap \mathbf{Z}^d - F \cap \mathbf{Z}^d$. In the course of the proof, an interesting result on the subsemigroups of $\mathbf{Z}^d$ (inside $\mathbf{R}^d$) is proved: If A is a subsemigroup of $\mathbf{Z}^d$ (under addition) such that $\mathbf{0}$ belongs to A and $A - A = \mathbf{Z}^d$, and b is an element of $\mathbf{Z}^d$ that is in the interior of the convex hull (defined using $\mathbf{R}^d$, of course) of A, then there exists an integer m so that n exceeding m entails that nb belong to A. The one-dimensional version of this is extremely well-known.

Chapter IV contains a proof that all finitely generated projective $R_K$-modules are free. This result is proved by means of what is occasionally called the "Legendre transformation" in convex analysis. There is a natural map associated to a positive Laurent polynomial P, $(\mathbf{R}^d)^{++} \longrightarrow \mathrm{Int}\, K$ which is a homeomorphism, and this extends to a homeomorphism between K and what is known as the pure state space of $R_K$. It follows that the latter is contractible, and then it is easy to see that all projectives are free. On the other hand, the $R_P$'s (from which the $R_K$'s are constructed) may have interesting Picard groups, and some examples are computed in Appendix A.

Section V is rather technical, relating some of the real geometry of the pure state space to the order structure of certain ideals.

Sections VI and VII concern a very special class of integral polytopes, which I have called integrally simple. They satisfy the integral analogue of simplicity for real polytopes (a polytope is simple if every proper face is of its dual is a simplex; dual polytopes of integral polytopes are not usually integral, so the definition does not involve duality). Then $R_K$ is factorial if and only if K is integrally simple. The length of section VI is due to the inclusion of several integro-geometric criteria for integral simplicity.

Section VII returns to the K-theory of fixed-point C* algebras. Specifically, in the case that K = cvx Log P is integrally simple, this section contains a description of all the primitive (two-sided) ideals in $A^T$; these are given via the projections that generate them. Of course, this is possible because there is a natural bijection between primitive ideals of $A^T$ and meet-irreducible order ideals of $R_K$. The work here is to describe all of the latter; it turns out that there is a pleasant geometric description of them. In the course of the discussion, we run across another result of the type described in section II, concerning eventual positivity of polynomials. Specifically, if Log P is obtained by stripping a layer corresponding to a facet (maximal face) of cvx Log Q, and the latter is integrally simple, then $P^n f$ being positive implies $Q^m f$ is as well for some m and n. This fails without integral simplicity.

Section VIII deals with automorphisms of and isomorphisms between the $R_K$'s and $R_P$'s. If K is either (real) indecomposable or integrally simple (the two properties are virtually disjoint) and K' is any integral polytope, then any ring isomorphism $R_K \longrightarrow R_{K'}$ is automatically an order-isomorphism. The description of automorphisms of $R_K$ is quite interesting and leads to an apparently difficult problem in algebraic geometry. After eliminating a finite top, known automorphisms consist of $(\mathbf{R}^d)^{++}$ acting by 'translation'; this appears to be all of them, and this is all of them if d = 1. On the other hand, if P is irreducible as an element of the Laurent polynomial algebra, $\mathbf{R}[x_i^{\pm 1}]$, then $R_P$ admits only finitely many order-automorphisms (that is, ring automorphisms for which both they and their inverses are order-preserving), all of which are uniquely determined by their action on K through AGL(d,Z) (in particular, the order-automorphisms of $R_P$ are faithfully represented as a permutation group on the vertices of K).

There is a notion of integral indecomposability for integral polytopes which is weaker than the usual one; unfortunately it is not generally preserved on replacing K by 2K. However, it turns out that the only integrally indecomposable integrally simple polytopes are multiples of the standard simplex (up to AGL(d,Z)-equivalence), and are therefore indecomposable in the usual sense.

As in section VII, §VIII contains an eventual positivity result for polynomials.

The first Appendix (A), contains computations of the Picard group of $R_P$, where cvx Log P is integrally simple. The Picard group is generated by an obvious set of prime ideals corresponding to the facets of K (a facet is a face of codimension one). For example, if P in $R[x_i^{\pm 1}]^+$ is irreducible (i.e., does not factor non-trivially), then $R_P$ has Picard group $Z^{e-d-1} \oplus Z/aZ$, where e is the number of facets, d is the dimension of cvx Log P, and a is a nonzero integer easily determined from Log P. By considering reducible polynomials, we obtain a result (possibly well-known in a more general context) about the number of non-homothetic indecomposables appearing as summands of multiples of K, for K integrally simple, specifically that K is expressible as a sum of the maximum number known to be possible, $e-d$. The proof involves a very simple dimension argument using the structure of $Pic(R_P)$.

Appendix B describes the connection between $A^T$ and $R_P$, in order that the result of (for example) section VII may be understood. Moreover, it also contains another, this time C*-algebraic, proof of a result in [H1; VIII] that the "obvious" ideals of $R_P$ or $R_K$ that correspond to faces actually are prime, and their quotients are what they appear to be.

The third appendix (C) deals with a variation on the original random walk problem. Here, we have a sequence $\{P_1, P_2, \ldots\}$ of polynomials with positive coefficients, with Log $P_i$ = Log P, for all i, for some fixed P in $R[x_i^{\pm 1}]^+$. With a suitable bounded below hypothesis on the nonzero coefficients of all of the $P_i$, we obtain, that if $P^n f$ has negative coefficients for some n, then there exists N so that $P_1 P_2 \ldots P_N f$ has the same property. In allowing P to change (somewhat), we are considering time-dependent (rather than time-independent) random walks.

Appendix D provides a large class of integrally simple polytopes arising from actions of Weyl groups of compact connected Lie groups (and some dihedral groups) on $Z^d$. With respect to a suitable sublattice of $Z^d$, the convex hull of almost all orbits of these actions on the dual of the maximal torus is integrally simple. In the case of the Weyl group of SU(d+1), precisely which orbits yield integrally simple polytopes is completely determined. It turns out that for SU(d+1), the integral polytope arising from an orbit is integrally simple if and only if it is simple (as a real polytope), but this fails for SO(5).

Finally, Appendix E contains an elementary proof that the Legendre transformation used in section IV is a homeomorphism. The only other proof I could locate in the literature [Ro; Theorem 26.5] uses a great deal of convex analysis, so would require a large investment of time on the part of the interested reader.

This was originally typed in 1985. Extensive revisions were made possible by using a manual optical scanner ("Omni-Reader" distributed by G.A.S. International) to read it as a text file into my Macintosh™. Formatting and revisions were done using the software Microsoft® Word 3.0 (upgraded from 1.05 halfway through). Most of the equations that are displayed, and some of those that are on-line were created with McΣqn, and some diagrams used MacDraw. The printing was done on an Apple® LaserWriter™.

This work was primarily supported by an operating grant from the Natural Sciences and Engineering Research Council (Canada), and a Steacie Memorial Fellowship. Helpful comments were received from Ken Goodearl, and a useful suggestion from Mel Hochster.

# I. DEFINITIONS AND NOTATION

Throughout, $\mathbf{Z}^d$ will denote the concrete group of lattice points in $\mathbf{R}^d$. If $w = (w(1),...,w(d))$ is a lattice point, we define the monomial in the Laurent polynomial ring $\mathbf{R}[x_i^{\pm 1}]$ (or $\mathbf{Z}[x_i^{\pm 1}]$) in the $d$ variables, $x_1,...,x_d$,

$$x^w = x_1^{w(1)} x_2^{w(2)} \dots x_d^{w(d)}.$$

If $P = \sum \lambda_w x^w$ is an element of $\mathbf{R}[x_i^{\pm 1}]$ (or of $\mathbf{Z}[x_i^{\pm 1}]$), that is, if $\lambda_w$ are all real numbers (integers), we define

$$\text{Log } P = \{w \in \mathbf{Z}^d \mid \lambda_w \neq 0\}.$$

If $S$ is a finite subset of $\mathbf{Z}^d$, cvx $S$ will denote the convex hull of $S$ with respect to $\mathbf{R}^d$; $d_e$cvx $S$ will denote the set of extreme points, that is, vertices, of cvx $S$. When $P$ belongs to $\mathbf{R}[x_i^{\pm 1}]$, we frequently write $d_e$Log $P$ for $d_e$ cvx Log $P$. Notice that $S$ always contains $d_e$cvx $S$.

Defining $\mathbf{R}[x^{\pm 1}]^+ = \{\sum \lambda_w x^w \mid \lambda_w$ are all non-negative real numbers$\}$ (similarly for $\mathbf{Z}[x_i^{\pm 1}]^+$), $\mathbf{R}[x_i^{\pm 1}]$ becomes a partially ordered ring having $\mathbf{R}[x^{\pm 1}]^+$ as its positive cone. We write (for $P$ and $Q$ in $\mathbf{R}[x_i^{\pm 1}]$), $P \leq Q$ in $\mathbf{R}[x_i^{\pm 1}]$ if all the coefficients of $Q-P$ are non-negative. Now let $P$ be a positive element of $\mathbf{R}[x_i^{\pm 1}]$, i.e., all of its coefficients are non-negative. Following the constructions in [H1; section I], we form the partially ordered ring $S_P = \mathbf{R}[x_i^{\pm 1}, P^{-1}]$ by inverting $P$ and taking the limit ordering, which works out to

$$S_P^+ = \{f/P^k \mid f \in \mathbf{R}[x_i^{\pm 1}] \text{ and there exists } N \text{ so that } P^N f \in \mathbf{R}[x_i^{\pm 1}]^+\}.$$

In a partially ordered abelian group $G$ (for example, as $S_P$ is, as an abelian group), an element $u$ is called an <u>order unit</u> (for $G$) if $u$ is positive and for all $g$ in $G$, there exists an integer $N$ so that $g \leq Nu$. An <u>order-ideal</u> of $G$ is a subgroup $H$ of $G$ such that $H = H \cap G^+ - H \cap G^+$ ($H$ is <u>directed</u>) and $0 \leq g \leq h$, $g \in G$, $h \in H$ imply $g$ belongs to $H$ ($H$ is <u>convex</u>). If $x$ belongs to the positive cone of $S$, we may form the order ideal generated by $x$,

$$\{g \in G \mid \text{there exists } N \text{ in } \mathbf{N} \text{ so that } -Nx \leq g \leq Nx\}.$$

It is straightforward to verify that this is an order ideal of $G$, and that $x$ is an order unit for it.

Define $R_P$ to be the order ideal of $S_P$ generated by $1$, as an ordered abelian group. Then $R_P$ is a partially ordered ring, $1$ is an order unit for $R_P$, and

$$R_P = \mathbf{R}[x^w/P;\ w \in \text{Log } P] \qquad R_P^+ = \langle x^w/P;\ w \in \text{Log } P\rangle$$

(the latter is the semigroup generated additively and multiplicatively over $\mathbf{R}^+$ by the set enclosed therein), and for f in $\mathbf{R}[x^{\pm 1}]$, $f/P^k$ belongs to $R_P$ if and only if there exists an integer N so that $\mathrm{Log}(P^N f)$ is contained in $\mathrm{Log}\, P^{N+k}$. For more details, see [H1; section I].

Let us now compute some examples. We shall refer back to these, or to variations on them, throughout this monograph.

**EXAMPLE 1A** [H1; §V]. This is the most basic case. Set $P = 1 + \sum_{1\le i\le d} x_i$. Then $\mathrm{Log}\, P = \{\mathbf{0}, (1,0,\ldots,0), (0,1,0,\ldots,0), \ldots, (0,\ldots,0,1)\}$ and cvx Log P is the standard d-simplex in $\mathbf{R}^d$, that is, $\{r = (r_i) \in \mathbf{R}^d \mid r_i \ge 0, \sum r_i = 1\}$, and we shall call this K. As a ring, $S_P$ is simply $\mathbf{R}[x_i^{\pm 1}, P^{-1}]$; the positive cone is more complicated.

Now $R_P = \mathbf{R}[x_i/P]$ (note that $1/P = 1 - \sum x_i/P$). Set $X_i = x_i/P$. Then it is easy to check that $\{X_i\}$ is algebraically independent, so that $R_P = \mathbf{R}[X_i]$, i.e., it is just a pure polynomial algebra, in the variables $X_i$. The positive cone is generated (multiplicatively and additively) by the set $\{X_1, X_2, \ldots, X_d; 1-\sum X_i\}$ (observe that the latter term is just $1/P$). In principal, the positive cone can be determined via [H2; Theorem B], but it is quite complicated.

If h in $\mathbf{R}[X_i]$ is in the positive cone, then h is a positive (real-) linear combination of products of terms in $\{X_1, X_2, \ldots, X_d; 1-\sum X_i\}$. It follows that viewed as a function on K,

(*) h is strictly positive on the interior of K, *and*
the zero set of h (intersected with K) is a union of faces of K.

Both statements are immediate consequences of the fact that each of the $X_i$ satisfy (*).

Unless d = 1, the conditions in (*) are not sufficient for an element h in $\mathbf{R}[X_i]$ to lie in the positive cone. However, it is true that if the first statement is strengthened to h being strictly positive on *all of* K, then h *is* in the positive cone [H4].∎

**EXAMPLE 1B** Set $P = \prod_{1\le i\le d} (1+x_i)$. Then Log P is the set of vertices of the standard cube in $\mathbf{R}^d$, and cvx Log P = K is the standard cube itself. From the definitions and the fact that in this case, $\mathrm{Log}\, P = K \cap \mathbf{Z}^d$ (which is not always true), $R_P = \mathbf{R}[x^w/P \mid w \in K \cap \mathbf{Z}^d]$. Using partial fractions, we see that if $X_i = x_i/(1+x_i)$, then again $R_P = \mathbf{R}[X_i]$, a pure polynomial algebra. With a little more care, it is possible to conclude that the positive cone is generated by $\{X_i; 1-X_i\}$.

Notice that here $K = \{r \in \mathbf{R}^d \mid X_i(r) \geq 0$ and $(1-X_i)(r) \geq 0$ for $i = 1,2,...,d\}$, and again that if $h$ is in the positive cone of $R_P$, but is viewed as an element of $\mathbf{R}[X_i]$, then (*) of example 1A (with $K$ and the $X_i$ having been translated to this present setting) holds.∎

In both of these examples, $R_P$ happened to be a unique factorization domain, even a pure polynomial algebra. We shall see in Appendix A that this is an extremely improbable event. Specifically, A.8A asserts that if $P$ is an element of $\mathbf{R}[x_i^{\pm 1}]^+$ and $P$ is irreducible (in the algebraic sense—this is generic, that is, a "random" selection of $P$ will almost always be irreducible), and an additional innocuous condition holds, and if $R_P$ admits unique factorization, then up to the natural action of $AGL(d,\mathbf{Z})$ on $\mathbf{Z}^d$, $P$ must be in a form very similar to that described in Example 1A, $P = \lambda_o + \sum \lambda_i x_i$ (the $\lambda$'s are strictly positive real numbers). In particular, when the hypotheses apply, cvx Log P must be a standard d-simplex. Of course, in Example 1B, $P$ is not irreducible (see also Example VIII.19). Now we give a more typical example:

**EXAMPLE 1C** Put $d = 2$, $x_1 = x$, and $x_2 = y$; set $P = 1 + x + y + y^2$. Then Log P is $\{(0,0),(1,0),(0,1),(0,2)\}$ and cvx Log P is right triangle of height 2 and width 1. Now $R_P = \mathbf{R}[x/P, y/P, y^2/P, 1/P]$. Set $X = 1/P$, $Y = y^2/P$, and $Z = y/P$. Then as rings, $R_P = \mathbf{R}[X, Y, Z]$ (as $x/P = 1-X-Y-Z$), and there is one relation, namely, $Z^2 = XY$. Let $\underline{X},\underline{Y},\underline{Z}$ be variables, so that $R_P$ is a quotient of $\mathbf{R}[\underline{X},\underline{Y},\underline{Z}]/(\underline{Z}^2 - \underline{X}\underline{Y})$. However, this is precisely the ring in [ZS; p. 154, para. 2], so is a domain, and as $R_P$ has transcendence degree 2, and any further relations would reduce the transcendence degree, we deduce that $R_P$ is $\mathbf{R}[\underline{X},\underline{Y},\underline{Z}]/(\underline{Z}^2 - \underline{X}\underline{Y})$.

The relation $Z^2 = XY$ gives two inequivalent factorizations into irreducibles (it will be shown in VIII.18, that if $P$ is irreducible in $\mathbf{R}[x_i^{\pm 1}]^+$ and $w$ belongs to Log P, then $x^w/P$ is an irreducible element of $R_P$). Thus $R_P$ is not a unique factorization domain (viz. op.cit.).∎

In examples 1A, 1B, there was an elementary transformation ($X_i = x_i/P$ in 1A, and $X_i = x_i/(1+x_i)$ in 1B), that converted elements of $R_P$ into functions, even polynomials, on $K =$ cvx Log P. In example 1C and indeed more generally, there is a similar transformation arising out of the Legendre transformation of convex analysis. This will be discussed in section IV. The functions on cvx Log P cannot be realized as polynomials in general, but the conditions in (*) will still hold for elements in the positive cone.

Now we revert to general remarks about $R_P$. Suppose that $P$ is in $\mathbf{R}[x_i^{\pm 1}]^+$; then $\mathrm{Log}\, P^a = a \,\mathrm{Log}\, P$ (all sums of $a$ elements of $\mathrm{Log}\, P$). Because $R_P$ inherits the relative ordering from $S_P$, if $f/P^k$ belongs to $R_P$, then it is positive (in $R_P$) if and only if there exists $N$ so that $P^N f$ has no negative coefficients. Inasmuch as $R_P$ is an order ideal in a direct limit of dimension groups (see [EHS]), $R_P$ is itself a dimension group, although the countability restriction will have to be omitted. This means that it satisfies:

(a) **Unperforation:** For all $g$ in $G$, if for some $m$ in $\mathbf{N}$, $mg \in G^+$, then $g \in G^+$.

(b) **Interpolation:** For all $a,b,c,d$ in $G$ with $a,b \le c,d$ (that is, $a \le c$; $b \le c$; $a \le d$; $b \le d$), then there exists $e$ in $G$ so that $a,c \le e \le c,d$.

An <u>integral polytope</u> $K$ in $\mathbf{R}^d$ is a compact convex set all of whose vertices are in $\mathbf{Z}^d$ (in particular, it has only finitely many vertices). Let $S$ be a finite subset of $\mathbf{Z}^d$; a special case occurs when $S = K \cap \mathbf{Z}^d$ for some integral polytope $K$. We define a partially ordered ring $R_S$ as follows.

Let $P = \sum_{w \subset S} \lambda_w x^w$ where the $\lambda_w$ are strictly positive real numbers. Define $U(R_P)$ or simply $U$ if there is likely to be no confusion, to be the set of all order units of $R_P$, and set $R_S = U^{-1}R_P$; in other words, we invert all the order units. Note that the set of order units is closed under multiplication; it is even closed under addition!

Now we define an ordering on $U^{-1}R_P$ by setting

$$R_S^+ = \{au^{-1} \mid a \in R_P, u \in U, \text{ and there exists } v \text{ in } U \text{ with } av \in R_P^+ \}.$$

There are three obvious problems that present themselves in this definition. The first is that, at first glance, $R_S$ could depend upon the choice of coefficients $\{\lambda_w\}$ for $P$. It does not, and the argument is fairly easy (I.1). The second is, that the ordering looks a bit awkward to work with. In fact it turns out that for $a$ in $R_P$:

$$\text{if } au^{-1} \in R_S^+, \text{ then already } a \in R_P^+. \qquad (\dagger)$$

This is a restatement of the main result of section II, and is equivalent to a random walk problem in $\mathbf{Z}^d$. The third apparent difficulty concerns the nature of order units in $R_P$; these have been described reasonably well in [H1; V.4 and V.5].

Assuming the first difficulty (independence of the choice of $\{\lambda_w\}_{w \in S}$) has been resolved, it becomes clear that $R_S$ is an $AGL(d,\mathbf{Z})$-invariant; if there exists $h$ in $AGL(d,\mathbf{Z})$ such that $hS = S'$, then $R_S$ is isomorphic (as a partially ordered ring) to $R_{S'}$.

**I.1 PROPOSITION.** Let P and Q be elements of $\mathbf{R}[x^{\pm 1}]^+$ such that $m \operatorname{Log} P = n \operatorname{Log} Q = pS$ for some integers m, n, p, and S a finite set of lattice points. Then we have that

$$U(R_P)^{-1} R_P = U(R_Q)^{-1} R_Q,$$

so that $R_S$ is defined independently of the choice of P.

Proof: Let $r = au^{-1}$ belong to the left side, with a in $R_P$ and u an order unit thereof. We may find k sufficiently large so that $a = f/P^k$, $u = g/P^k$, with $k \operatorname{Log} P \supset \operatorname{Log} f$, $k \operatorname{Log} P \supset \operatorname{Log} g$, and g having only non-negative coefficients. By [H1; V.4], $d_e \operatorname{Log} g = d_e \operatorname{Log} P^k = \{kv \mid v \in d_e \operatorname{Log} P\}$. So $r = f/g$.

Consider $t = Q^N/P^M$; this certainly belongs to $R_P^+$, and by [H1;V.5], it is an order unit. Select positive integers c,e so that $k + c = em$. Then

$$r = \frac{f}{g} = \frac{fP^c/P^{em}}{gP^c/P^{em}} = \frac{fP^c/Q^{en}}{gP^c/Q^{en}}.$$

With $a' = fP^c/Q^{en}$, and $u' = gP^c/Q^{en}$, we will show that a',u' belong to $R_Q$, and that u' is an order unit. To see this, consider

$$\operatorname{Log} fP^c \subset \operatorname{Log} f + c \operatorname{Log} P \subset (k+c) \operatorname{Log} P = em \operatorname{Log} P = en \operatorname{Log} Q = \operatorname{Log} Q^{en},$$

and similarly $\operatorname{Log} gP^c \subset en \operatorname{Log} Q$, so that a',u' belong to $R_Q$. Now $gP^c$ has no negative coefficients, so that u' lies in $R_Q^+$. As g and $P^c$ have only positive coefficients and $\operatorname{cvx} \operatorname{Log} g = k \operatorname{cvx} \operatorname{Log} P$, it follows that

$$\operatorname{cvx} \operatorname{Log} P^c g = (k+c) \operatorname{cvx} \operatorname{Log} P = em \operatorname{cvx} \operatorname{Log} P = en \operatorname{cvx} \operatorname{Log} Q.$$

Hence $d_e \operatorname{Log} P^c g = d_e \operatorname{Log} Q^{en}$, so by [H1; V.5], $u' = P^c g/Q^{en}$ is an order unit. As $r = a'(u')^{-1}$ belongs to $U(R_Q)^{-1} R_Q$, we have one inclusion of rings; by symmetry, we obtain equality.∎

A (normalized) state of a partially ordered abelian group is an order-preserving group homomorphism $f: G \longrightarrow \mathbf{R}$ such that $f(u) = 1$. A state is pure if it cannot be expressed as a non-trivial convex linear combination of distinct states. We quote several results that will be used repeatedly without specific reference:

> If (G,u) is an unperforated partially ordered abelian group and g is an element of G such that $\alpha(g) > 0$ for all pure states $\alpha$ of (G,u), then $g \in G^+$ [H1; I.1].

If $R$ is a partially ordered commutative ring having 1 as an order unit, then all pure states are multiplicative and any state that is multiplicative is pure [H1;I.2(b)].

This allows us to obtain a clearer picture of the set, $U$, of order units of $R_P$. As a consequence of these facts, an element $f/P^k$ of $R_P$ is an order unit if and only if $\alpha(f/P^k) > 0$ for all pure states $\alpha$ of $R_P$ (this entails that $\{\alpha(f/P^k) \mid \alpha$ varying over the pure states of $R_P\}$ be bounded below). The pure states of $R_P$ have been determined in [H1; §III]; we now describe a dense set of them.

Let $r = (r_1,r_2,...r_d)$ be a member of $(\mathbf{R}^d)^{++}$; that is, the $r_i$ are all strictly greater than zero. We attach a pure state to $r$, called the <u>point evaluation at r</u> defined via

$$\alpha_r(f/P^k) = f(r)/P^k(r);$$

this is simply point evaluation at $r$. Endowed with the point-open topology (i.e., as a subset of $\mathbf{R}^{R_P}$) the set $\{\alpha_r \mid r \in (\mathbf{R}^d)^{++}\}$ admits the same topology as the usual one when it is identified in the obvious way with $(\mathbf{R}^d)^{++}$. By [H1; III.3], this set of point evaluations is dense in the set of pure states of $R_P$. Hence, provided that $f/P^k$ belongs to $R_P$, the former is an order unit if and only if it is bounded below (away from zero) as a function on $(\mathbf{R}^d)^{++}$. The rational functions that satisfy these conditions have been determined in [H1; V.4].

Specifically, $f/P^k$ is an order unit of $R_P$ if and only if:

$\text{cvx Log } f = \text{cvx Log } P^k$;

there exists $m$ so that $\text{Log } f + \text{Log } P^m \subset \text{Log } P^{m+k}$;

$f|(\mathbf{R}^d)^{++} > 0$

for every proper face $F$ of $\text{cvx Log } P^k$, $f_F|(\mathbf{R}^d)^{++} > 0$;

here we use the notation $f_F$ to represent the polynomial $\sum_{w \in F \cap \mathbf{Z}^d} \mu_w x^w$ where $f = \sum \mu_w x^w$. If $f/P^k$ is an order unit, then of course, there exists an integer $n$ so that $P^n f$ has no negative coefficients. Then $f/P^k = fP^n/P^{k+n}$, and now the numerator has no negative coefficients. If we assume that $f$ has no negaitive coefficients, then $f/P^k$ is an order unit if and only if the first two conditions hold. We now apply these definitions to the basic examples 1A, 1B, and 1C.

EXAMPLE 1A (revisited). Here every order unit of $R_P$ can be written in the form $g/P^k$ where $g$ has only positive coefficients, each of $x^0, x_1^k, x_2^k, ..., x_d^k$ appears (i.e., with nonzero hence positive coefficient) in $g$, and such that the total degree of $g$ is $k$. Thus

$$R_P[U^{-1}] = \{f/g \mid f \in \mathbf{R}[x_i^{\pm 1}],\ g \in \mathbf{R}[x_i^{\pm 1}]^+,\ \mathrm{Log}\, f \subset \mathrm{Log}\, g, \text{ and } \mathrm{cvx}\, \mathrm{Log}\, g = kK \text{ for some integer } k\}.$$

(Recall that in the context of Example 1A, $K$ is the standard $d$-simplex.) To see this, simply write $f/g = (f/P^k)\cdot(g/P^k)^{-1}$ and note that $\mathrm{Log}\, P^k = k\,\mathrm{Log}\, P = kK \cap \mathbf{Z}^d$ in this particular example (for general $P$, it is not usually true that $k(\mathrm{Log}\, P) = k(\mathrm{cvx}\, \mathrm{Log}\, P \cap \mathbf{Z}^d)$).

If we now apply the change of variables, $X_i = x_i/P$. then the order units become precisely the elements $h$ of $\mathbf{R}[X_i]$ that are strictly positive on $K$ (as follows from the criteria stated above; see also [H4]). So in this case, $R_P[U^{-1}]$ has the particularly elementary description:

$$\{h/h' \mid h, h' \in \mathbf{R}[X_i] \text{ and } h'|K > 0\}.$$

In particular the transformation $X_i = x_i/P$ actually induces a homeomorphism from the pure state space of $R_P$ to $K$. This is what happens in general (§IV), but the formula is more complicated.∎

The behaviour of Example 1B is very similar. In the case of Example 1C, we describe the map from the pure state space of $R_P$ to $K$. There is no change of variables possible, but the map on the pure states is described by (§IV):

$$\gamma \mapsto \gamma(x/P)(1,0) + \gamma(y/P)(0,1) + \gamma(y^2/P)(0,2) \in K,$$

for a pure state $\gamma$. Notice that the point evaluations $\gamma = \alpha_r$ are mapped to the interior of $K$. In the case of Example 1A, the corresponding map is especially simple:

$$\gamma \mapsto \sum \gamma(x_i/P)\, e_i$$

where $e_i$ is the i-th standard basis vector of $\mathbf{R}^d$. When restricted to the point evaluations which are identified with $(\mathbf{R}^d)^{++}$, the map $(\mathbf{R}^d)^{++} \longrightarrow K$ is given by:

$$r = (r_1, r_2, \ldots r_d) \mapsto \frac{1}{1+\sum r_i}(r_1, r_2, \ldots r_d),$$

which is clearly a homeomorphism $(\mathbf{R}^d)^{++} \longrightarrow \mathrm{Int}\, K$.∎

If $K$ is an integral polytope, we write $R_K$ for $R_{K\cap \mathbf{Z}^d}$. Very often, we assume that $P$, $K$, or $S$ are projectively faithful; we say that $P$ is projectively faithful if $\mathrm{Log}\, P - \mathrm{Log}\, P$ generates all of $\mathbf{Z}^d$ as an abelian group; $S$ is projectively faithful if $S - S$ generates $\mathbf{Z}^d$, and $K$ is if the set $S = K \cap \mathbf{Z}^d$ is.

When K is not projectively faithful, we can usually reduce to a smaller polytope which is; we do this by finding a **Z**-basis for the group generated by $K \cap \mathbf{Z}^d - K \cap \mathbf{Z}^d$, relabelling everything according to this basis, obtaining a new $\mathbf{Z}^{d'}$ and $\mathbf{R}^{d'}$ (with $d' < d$ a possibility). Then $R_K, R_P, R_S$ will be unaffected (up to order-isomorphism) by this process. These definitions are motivated by results about representations of tori and other compact groups, see [H1; section II.1].

We have defined $R_P, R_K, R_S$ using real coefficients. It is sometimes useful to be able deal with their integer analogues (which means rational coefficients for the latter two). Thus if P belongs to $\mathbf{Z}[x_i^{\pm 1}]^+$, define $S_{P,\mathbf{Z}}$ as $\mathbf{Z}[x_i^{\pm 1}, P^{-1}]$, and analogously $R_{P,\mathbf{Z}}$. All results for the real versions go over to the integer versions, except as noted. However, $\dim_{\mathbf{R}} R_S$ is uncountable, but $\dim_{\mathbf{Q}} R_{S,\mathbf{Z}}$ is countable; thus while $R_P = R_{P,\mathbf{Z}} \otimes \mathbf{R}$, $R_S$ is not isomorphic to $R_{S,\mathbf{Z}} \otimes \mathbf{R}$. Where the translation is correct but not completely obvious, we shall make a brief comment.

## II. A RANDOM WALK PROBLEM

Suppose one is given a (finite) initial distribution of masses, f, (allowing negative masses) on a lattice $\mathbf{Z}^d$, together with two processes P and Q (see illustration 2.1, for example).

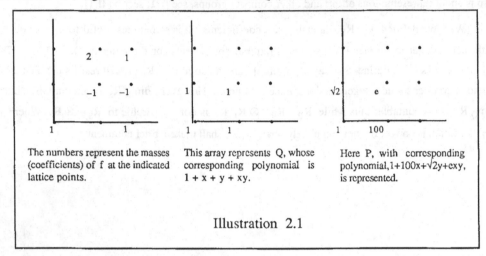

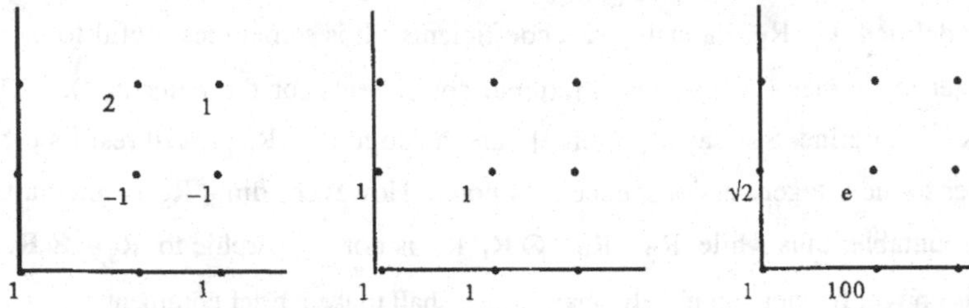


The numbers represent the masses (coefficients) of f at the indicated lattice points.

This array represents Q, whose corresponding polynomial is $1 + x + y + xy$.

Here P, with corresponding polynomial, $1+100x+\sqrt{2}y+exy$, is represented.

Illustration 2.1

The processes apply to f by convolution, i.e., by applying to each point in f, and summing the resulting masses at each point, to create a new distribution. Assume that P and Q have identical support (as in the illustrated example) but with possibly different masses, and that some iteration of Q will change f so that all of its point masses become non-negative. Does it follow that some iteration of P to the initial distribution f, will render the masses non-negative? Inasmuch as P is distorted in one direction, the reader might feel that the answer should be negative.

However, the answer is yes; it comes from a translation of Theorem II.1 (although the statement of the theorem contains a bit more information). In the example above, two iterations of Q are required to render the distribution at left non-negative, so that some very large (at least 100/e) number of iterations of P will do the same. This may be difficult to verify empirically, as the coefficients become astronomically large.

Of course, this random walk problem is equivalent to the question (†) discussed in section I.

The example translates as: (with $x = x_1$, $y = x_2$)

$$f = 1 + x^2 - xy - x^2y + 2xy^2 + x^2y^2$$

$$P = 1 + x + y + xy$$

$$Q = 1 + 100x + 2^{\frac{1}{2}} y + exy$$

and $P^2 f$ has only non-negative coefficients.

II.1 **THEOREM** Let P and Q be polynomials in $\mathbf{R}[x_i^{\pm 1}]^+$ $(1 \leq i \leq d)$ such that

$$d_e \mathrm{Log}\, P \subset \mathrm{Log}\, Q \subset \mathrm{Log}\, P.$$

Let f be a polynomial with real coefficients, such that $Q^n f$ has only non-negative coefficients for some integer n. Then there exists an integer N such that $P^N f$ has only non-negative coefficients.

If there is no obvious relation between the Log sets of P and Q, then there need be no obvious connection between then f's that become positive with respect to P, Q, respectively. For example, if $P = 1 + x + y$ (so $d = 2$, and cvx Log P is the solid triangle with vertices (0,0), (0,1), (1,0)) and $Q = 1 + x + y + xy$ (so $d = 2$ and cvx Log Q is the solid square with vertices (0,0), (0,1), (1,0), (1,1)), then there exist real polynomials in two variables, $f_o$, $f_1$, such that each of $P^n f_o$, $Q^n f_1$ have only non-negative coefficients for some n, while neither of $P^m f_1$, $Q^m f_o$ have only positive coefficients for *any* m (this example is given at the conclusion of the proof of II.1).

In order to prove the theorem, we actually prove that $R = R_P$, the bounded subring of $S = S_P = \mathbf{R}[x_i^{\pm 1}, P^{-1}]$ (see section I), satisfies:

If u is an order unit of R, and a is an element of R, with ua positive in R,

then a is positive in R.

An order-ideal I in a partially ordered group is <u>meet-irreducible</u> (formerly: <u>prime</u>; see [EHS; section I] for an explanation of why it should be called prime, and [H1; section VII] for an explanation of why it should not be) if for order-ideals J and K, $I = J \cap K$ implies $I = J$ or $I = K$. Suppose now that I is an order ideal in a partially ordered ring R having 1 as an order unit. Then I is <u>order-primary</u> if for a, b in $R^+$,

$$a \notin I \text{ and } ab \in I \text{ jointly imply that } b^n \in I \text{ for some } n.$$

(That is, in the quotient ordered ring, positive zero divisors are nilpotent.) An ideal of R is <u>primary</u> [ZS; p. 152, 204] if the hypotheses "a, b in $R^+$" are replaced by "a, b in R"; so zero divisors are nilpotent in the quotient ring.

We will show (in order to establish the order unit result cited above) that in the rings $R = R_P$ arising as bounded subrings from polynomial multiplications as above, meet-irreducible order-ideals are order-primary and

order-primary order-ideals are primary.

All results in this section then follow immediately.

The first implication is proved in II.3, and the argument adapts the analogous ring-theoretic result [ZS; p. 209, Lemma 2].

Define for this section only, a <u>monomial</u> in $R_P$ (of level k) to be a term of the form $x^w/P^k$ with w in $\mathrm{Log}\, P^k$.

**II.2 LEMMA** If r is an element of $R_P$ with

$$-Nx^w/P^k \leq r \leq Nx^w/P^k$$

for some positive integer N and some monomial $x^w/P^k$ in $R_P$ (that is, w belongs to $\mathrm{Log}\, P^k$), then the element of $S_P$ defined by $s = P^k x^{-w} r$ belongs to $R_P$, and so r lies in the principal ideal, $(x^w/P^k)R_P$.

<u>Proof</u>: Since $P^k$ is a positive element of $S_P$, so is $P^k x^{-w} = s_o$. Hence multiplication by $s_o$ is order-preserving. Thus,

$$-N\cdot 1 \leq r s_o = s \leq N\cdot 1.$$

By definition, s belongs to $R_P$ and obviously $r = s(x^w/P^k)$.∎

**II.2A PROPOSITION** Any ideal of $R_P$ generated by terms of the form $x^w/P^k$ (subject to $w \in \mathbf{Z}^d$, $k \in \mathbf{Z}^d$, and $w \in \mathrm{Log}\, P^k$) is an order ideal.

<u>Proof</u>: Let $\{e_i\}$ be a collection of monomials that generate the ideal I; suppose $s = \sum e_i a_i$ with $a_i$ in $R_P$ ($a_i = 0$ for almost all i). Since 1 is an order unit for $R_P$, there exists a positive integer N so that $-N \leq a_i \leq N$ for all the i for which $a_i \neq 0$. Thus

$$-N\sum_{a_i \neq 0} e_i \leq s \leq N\sum_{a_i \neq 0} e_i$$

so that I is directed. Suppose $0 \leq s \leq b$, with b in I and s in $R_P$. Then $0 \leq s \leq N\sum e_i$ for some finite set of monomials $e_i$ in I (since b belongs to I). By Riesz interpolation, there exist $c_i$ in $R_P$ such that $s = \sum c_i$ and $0 \leq c_i \leq Ne_i$. By 2.2, each $c_i$ belongs to I, so that s also belongs.∎

We note that $R_P$ is a finitely generated **R**-algebra (or ring, if the original coefficients were taken from **Z**), so is noetherian.

**II.3 PROPOSITION** If $I$ is an order-ideal of $R_P$ that is meet-irreducible (as an order-ideal), then $I$ is order-primary.

Proof: Suppose $a$ and $b$ are positive elements of $R_P$ with the product $ab$ lying in $I$, but neither $a$ belongs to $I$, nor does any power of $b$. Since each of $a, b$ is a positive sum of monomials (a product of monomials is obviously a monomial, possibly at a different level), we may find monomials $x^u/P^k \leq Na$ and $x^v/P^m \leq Nb$, for some integer $N$, so that $x^u/P^k$ does not belong to $I$, nor does any power of $x^v/P^m$. Since order ideals are convex, it easily follows that $(x^u/P^k)\cdot(x^v/P^m)$ belongs to $I$ (the product is dominated by $N^2ab$). Hence we may assume that $a$ and $b$ are actually monomials. Now we can mimic the proof of [ZS; p. 209, Lemma 2].

By Noetherianness of $R_P$, there exists an integer $k$ so that

$$(I{:}b^k) = (I{:}b^{k+1}) = (I{:}b^{k+2}) = \ldots$$

where $(I{:}c) = \{r = R_P \mid rc \in I\}$. Let $J$ be the order-ideal generated by $I$ and $b^k$, that is,

$$J = \{r \in R_P \mid \text{there exists a positive integer } N\text{, and } i \text{ in } I^+ \text{ with } -N(i+b^k) \leq r \leq N(i+b^k)\}.$$

Define $K$ to be the order-ideal generated by $I$ and $a$, that is,

$$K = \{r \in R_P \mid \text{there exists a positive integer } N\text{, and } i \text{ in } I^+ \text{ with } -N(i+a) \leq r \leq N(i+a)\}.$$

(Actually $J$ and $K$ are just the ***ideals*** generated by $I$ and $b^k$, respectively $I$ and $a$, because $a$ and $b$ are monomials, but we do not require this.)

We shall show that $I = J \cap K$. Certainly $I \subset J \cap K$; since $R_P$ has the Riesz interpolation property, $J \cap K$ is an order-ideal, and it thus suffices to show $(J \cap K)^+ \subset I$. Select $z$ in $(J \cap K)^+$. Then $0 \leq z \leq N(i+b^k)$ for some positive $i$ in $I$. By interpolation, we may write $z = i_o + b_o$ where $0 \leq i_o \leq Ni$ (so $i_o$ lies in I), and $0 \leq b_o \leq Nb^k$. By II.2 and the assumption that $b^k$ is a monomial, there exists $r_0$ in $R_P$ with $b_o = b^k r_0$.

We also have $0 \leq z \leq M(i+a)$ for some positive integer $M$; we may thus write $z = i_1 + a_o$, where $0 \leq i_1 \leq Mi$ (so $i_1$ lies in I) and $0 \leq a_o \leq Ma$. Again by I.2 (and recalling that $a$ is a monomial), there exists $r_1$ in $R_P$ with $a_o = ar_1$.

Hence $z = i_o + b^k r_0 = i_1 + ar_1$, with $i_o$ and $i_1$ in I. As $ab$ belongs to I, it follows that $b^{k+1}r_0$ belongs to I. Our hypothesis on $k$ yields that $b^k r_0$ lies in I, so that $z$ belongs to I.

As $I = J \cap K$ and I is meet-irreducible, either $I = J$ or $I = K$. The former entails that $b^k$ belong to I, the latter that a belong to I, and both are contradictions.∎

There exist commutative noetherian domains that are partially ordered rings and dimension groups for which the conclusion above fails. These arise from orthogonal functions [H5], as follows:

**EXAMPLE** Let $R_0 = \mathbf{R}[X]$, the polynomial ring in one variable, and define $u_i$ in $\mathbf{R}[X]$ recursively, via:

$$u_o = 1;\ u_1 = X;\ \text{and for } i \geq 1,\ u_{i+1} = (X - a_i)u_i - b_i u_{i-1},$$

$$\text{where } 0 < a_1 < a_2 \leq a_3 \leq a_4 \leq \ldots, \text{ and } 0 = b_1 \leq b_2 \leq b_3 \leq b_4 \leq \ldots .$$

As the sequences $\{a_i\}$ and $\{b_i\}$ are monotone non-decreasing, the positive cone generated (as an $\mathbf{R}^+$-semigroup) by $\{u_i\}$ is closed under multiplication [AI; 2.9], that is, for all j, k, $u_j u_k$ is a positive real linear combination of $\{u_i\}$. Thus, with the positive cone $R_o^+ = \{\sum \lambda_i u_i \mid \lambda_i \in \mathbf{R}^+, \lambda_i = 0 \text{ for almost all } i\}$, $R_0$ becomes a partially ordered ring. Set $u = 1+X$, and let S be $R_0[u^{-1}]$, equipped with the limit ordering with positive cone $R^+ = \{fu^{-k} \mid f \in \mathbf{R}[X], k \geq 0$, and there exists an integer n so that $fu^n \in R_o^+\}$. Now define R as the order ideal of S generated by 1 (as we have defined $R_P$ from $S_P$). As $R_0$ is clearly a dimension group, and S is obtained as the direct limit, $\lim R_0 \longrightarrow R_0$ with maps given by repeated multiplication by u, we see that R is a dimension group. It is straightforward to verify that R is generated as a real algebra by $Z = X/(1+X)$, so that $R = \mathbf{R}[Z]$ is just the pure polynomial algebra in the variable Z.

Define the maps $\beta_i : R \longrightarrow \mathbf{R}$ via $X \mapsto a_i$ ($a_o = 0$, so $\beta_i(f) = f(0), f(a_1)$ respectively for $i = 0, 1$). We can check that $\beta_i$ are order preserving (for $i = 0$ or $1$), and being multiplicative, are therefore pure states of R; define $I = \ker \beta_0$; this is an order ideal, and it is immediate that I is simply the ideal generated by Z, and is thus maximal. However, $\ker \beta_0 \cap R^+ \subset \ker \beta_1 \cap R^+$, so that $\ker \beta_1$ is not an order ideal. Specifically, let $a = a_1$, and set $t = a/(1+a)$, so $\beta_1(Z) = t$. Then $\ker \beta_0 = ZR$, and $\ker \beta_1 = (Z-t)R$. We claim that $(\ker \beta_1 \cap R^+)R = Z(Z-t)R$.

Since $Z(Z-t)$ is $(1+a)^{-1} u_2/u^2$, one inclusion is clear. One checks immediately that neither of $\pm(z-t)$ can be positive in R, since the value of $z-t$ at $\beta_0$ is $-t < 0$, while its value at the state given by $Z \mapsto 1$ (corresponding to $X \mapsto \infty$) is $1-t > 0$. As $Z(Z-t)R$ is contained properly in precisely the two ideals ZR, and $(Z-t)R$ (as R is just the polynomial algebra in Z), we deduce immediately that either $(\ker \beta_1 \cap R^+)R = Z(Z-t)R$ or $(\ker \beta_1 \cap R^+)R = (Z-t)R$. However, the former is not an order ideal, as $Z-t$ cannot be expressed as a difference of elements of $(Z-t)R$ (an elementary exercise). In particular, we deduce that $Z(Z-t)R$ is properly contained in only one proper order ideal, so is obviously meet-irreducible as an order ideal, but it is clearly not meet-irreducible as an ideal; similarly, it is order-primary, but not primary.∎

Before getting on with the proof of the theorem, we discuss order ideals in the ring $R_P$. It is an immediate consequence of II.2A that every order-ideal is generated ***as an ideal*** by monomials. Moreover, if $r$ is an element of the order-ideal $I$, then $r$ can be written in the form:

$$r = \sum_{w \in \operatorname{Log} P^k} \rho_w \frac{x^w}{P^k} \qquad \text{for some } k$$

such that every $x^w/P^k$ for which $\rho_w$ is not zero ***belongs to*** $I$.

To see this, write $r = b - c$ with $b$ and $c$ in $I^+$ (this is possible since I is directed). Write $b$ and $c$ as positive sums of monomials; by replacing, if necessary, $x^w/P^k$ terms by $x^wP/P^{k+1}$, and re-expressing this as a positive combination of monomials, we write for $k$ large enough,

$$b = \sum \mu_w\, x^w/P^k \qquad \text{and} \qquad c = \sum \sigma_v\, x^v/P^k,$$

with all the real coefficients $\mu_w, \sigma_v$ being strictly positive. Since $I$ is convex, all of the monomials occurring in these sums must belong to $I$, and the result (for $r$) follows.

Now we prove that if the order-ideal $I$ is order-primary, then it is primary as an ideal.

Suppose $e$ and $a$ are elements of $R_P$, with $a$ not in $I$ and $e^n$ not in $I$ for any $n$, but with the product $ea$ belonging to $I$. We wish to obtain a contradiction to $I$ being order-primary. Write $e = B/P^s$, $a = A/P^t$, with $A$ and $B$ in $\mathbf{R}[x_i^{\pm 1}]$, and $\operatorname{Log} B \subset \operatorname{Log} P^s$, $\operatorname{Log} A \subset \operatorname{Log} P^t$ (this can always be arranged, as $R_P$ is generated by monomials $x^w/P$, where $w$ lies in $\operatorname{Log} P$).

Now we make several reductions and improvements on the forms of A and B. Since $R_P/I$ is a commutative noetherian ring, it has an index of nilpotence; that is, there exists $n$ so that $r^m \in I$ implies $r^n \in I$. This $n$ will be fixed throughout the argument, and the statement "$r^n$ does not belong to I" will often be used as an abbreviation of "$r^m$ does not belong to $I$ for all $m$".

Write $B = \sum \rho_w x^w$. Set $W = \{w \in \operatorname{Log} B \mid (x^w/P^s)^n \in I\}$, and set $B_1 = B - \sum_{w \in W} \rho_w x^w$, and $e_1 = B_1/P^s$. Since a sum of nilpotents is nilpotent, $(e-e_1)^n$ belongs to I. Now $e_1$ does not belong to I, for if it did, both $e-e_1$ and $e_1$ would belong to the prime radical of I, and hence so would e, whence $e^n$ would belong to I, which it does not. Hence no power of $e_1^n$ belongs to I. Moreover $ae_1^n$ belongs to I, as

$$ae_1^n = (e-(e-e_1))^n a = e\,a(\text{stuff}) + (e-e_1)^n(\text{stuff}).$$

We would like to replace e by $e_1^n$, because of some improvements in the numerator. We have that $e_1^n = B_1^n/P^{ns}$. Suppose w lies in Log $B_1$; then we may write $w = \sum_{1\le i\le n} u(i)$, with u(i) in (Log B)\W, that is, $(x^{u(i)}/P^s)^n \notin I$. If $(x^w/P^{sn})^n$ belonged to I (that is, if some power of $x^w/P^{sn}$ belonged to I), then the product

$$\prod_{i=1}^{n}\left(\frac{x^{u(i)}}{P^s}\right)^n$$

would belong to I. Since $(x^{u(1)}/P^s)^m \notin I$ for all m, and since I is order-primary, it must be that

$$\prod_{i=2}^{n}\left(\frac{x^{u(i)}}{P^s}\right)^{n^2}$$

belongs to I; again, no power of $x^{u(2)}/P^s$ belongs to I, so from I being order-primary, some power of the product starting with i = 3 lies in I. This can obviously be continued until a contradiction is reached. The outcome of this is that for all w in Log $B_1^n$, no power of $x^w/P^{sn}$ lies in I. Replace B by $B_1^n$ (or rather, rename $B_1^n$, B), so that we have the following situation:

$$e = B/P^s,\ \text{Log } B \subset \text{Log } P^s,\ ae \in I,\ a \notin I,\ e^m \notin I \text{ for all } m,$$

AND (1)

$$w \in \text{Log } B \text{ implies no power of } x^w/P^s \text{ lies in } I.$$

The last statement is the improvement we wanted in B.

Now we improve A (without changing a,e, or B). We have that $ae = AB/P^{s+t}$. By the comment following II.3, there exists an element C in $\mathbf{R}[x_i^{\pm 1}]$ with Log C $\subset$ Log $P^\theta$ (for some integer $\theta$), such that $ae = C/P^\theta$ and for all w in Log C, $x^w/P^\theta$ lies in I (since ae belongs to I). Notice that $CP/P^{\theta+1}$ has the same properties; we may thus assume that $\theta \ge s+t$. Equating $AB/P^{s+t} = C/P^\theta$, we deduce an equality of polynomials (allowing negative exponents; this is not serious), $P^{\theta-(s+t)}AB = C$. Rename $P^{\theta-(s+t)}A$, as A, and $\theta-s$, t. This does not change a (which in the previous notation was $P^{\theta-(s+t)}A/P^{\theta-s}$), nor does it affect B or e. Hence, we may assume in addition to (1):

$a = A/P^t$, $e = B/P^s$

AND (2)

$w \in \text{Log}\, AB$ implies $x^w/P^{s+t}$ lies in I.

With these improvements, we can conclude the proof.

Define $V_A = \{w \in \text{Log}\, A \mid x^w/P^t \notin I\}$. If $A = \sum \lambda_w x^w$, set $\underline{A} = \sum \lambda_w x^w$, the sum taken over w in $V_A$. Suppose u lies in Log AB. Then there exists w in $V_A$ and v in Log B such that $u = w+v$, and so

$$x^u/P^{s+t} = (x^w/P^t)\cdot(x^v/P^s),$$

since $x^w/P^t$ does not lie in I, and neither does any power of $x^v/P^s$ (by (1)), it follows from I being order-primary that $x^u/P^{s+t}$ does not belong to I.

On the other hand, for all v' in Log AB, $x^{v'}/P^{s+t}$ belongs to I by (2). Thus $x^u$ cannot appear (with nonzero coefficient) in AB, and therefore, it must appear with nonzero coefficient in $(A-\underline{A})B$ (since it appears in $\underline{A}B$). Hence we may write $u = w'+v''$, where w' belongs to $(\text{Log}\, A)\setminus V_A$ and v'' lies in Log B. This yields

$$\frac{x^u}{P^{s+t}} = \frac{x^{w'}}{P^t}\cdot\frac{x^{v''}}{P^s}$$

which latter belongs to I (since w' lies in $(\text{Log}\, A)\setminus V_A$), a contradiction. This yields that $\underline{A}B = 0$, so either $\underline{A}$ or B is zero; thus either a belongs to I, or $e = 0$, and both are contradictions, concluding the proof.∎

We have shown:

II.4 **PROPOSITION** Any order-ideal of $R_P$ that is order-primary, is primary.

II.5 **THEOREM** Let P be a polynomial (allowing negative exponents) with only non-negative coefficients, and form $S_P = R[x_i^{\pm 1}, P^{-1}]$, and $R_P$ its bounded subring, with the ordering as defined earlier. Suppose u is an order unit of $R_P$, and a is an element of $R_P$ such that the product ua is positive (in $R_P$). Then a is positive in $R_P$.

<u>Proof:</u> Let I denote the order-ideal of $R_P$ generated by ua, that is,

$$I = \{r \in R_P \mid \text{there exists a positive integer } N \text{ with } -Nua \le r \le Nua\}$$

(this uses the positivity of ua); ua is of course an order unit for I.

If a+I is positive or zero in $R_p/I$, we may write $a = b-c$ where b lies in $R_p^+$ and c belongs to I. Then $ub = ua + uc$, so ub belongs to I. From the existence of some integer M so that $1 \leq Mu$, and the positivity of b, we have that $b \leq Mub$; thus b belongs to I, so a belongs to I. Let $\theta$ be a pure state of the ordered group with order unit (I, ua) (see [H1;section I]). By [H1; I.2], there exists a pure state L of $R_p$ such that $\theta(ua) = L(u)\theta(a)$. Since u is positive, L(u) is non-negative, and as $\theta(ua) = 1$, $\theta(a) > 0$. By [EHS; I.4] or [H1; I.1], a is positive.

This leaves the possibility that a+I is not positive in $R_p/I$. Since $R_p/I$ has the Riesz interpolation property, there exists a meet-irreducible order ideal, J, containing I, such that a+J is not positive or zero in $R_p/J$ (apply Zorn's Lemma to $R_p/I$; the existence of such an order-ideal is equivalent to the existence of sufficiently many primitive quotients of AF C*-algebras, and was observed by Fuchs in [F]). Now u does not belong to J (being an order unit), and neither does a (by hypothesis); however, ua belongs to I, which is contained in J. Hence u, a are non-trivial zero divisors modulo J; by I.2 and I.3, u must be nilpotent modulo J. Hence $u^n$ lies in J for some n. However, $1 \leq Mu$ entails

$$1 \leq Mu \leq M^2u^2 \leq \ldots \leq M^n u^n,$$

so $u^n$ is an order unit, and thus does not belong to J, a contradiction.∎

The proof above requires only that in the partially ordered ring R (having interpolation), meet-irreducible order ideals be primary as ideals. The example after I.3 admits a meet-irreducible order ideal that is not primary; on the other hand, even in that example, the "order-cancellation property" described above still holds.

Proof of Theorem II.1: By translation, we may assume 0 belongs to both Log P and Log f. Let Y be the subgroup of $\mathbf{Z}^d$ (the standard sublattice of $\mathbf{R}^d$) generated by Log f, and let Z be the subgroup of $\mathbf{Z}^d$ generated by Log P.

First suppose that $Y \subset Z$. Then we claim there exists w in $\mathbf{Z}^d$ such that $w \in \mathrm{Log}\, f \subset \mathrm{Log}\, P^m$ for some large integer m (observe since all the coefficients of P are non-negative, $\mathrm{Log}\, P^m$ is just $m\mathrm{Log}\, P$, the set of sums of m terms from Log P). To see this, find a $\mathbf{Z}$-basis for Z inside Log P, $\{w_1,\ldots,w_e\} \subset \mathrm{Log}\, Q$. Given v in Log f, there exists an e-tuple of integers $\{\lambda_{vi}\}$ such that $v = \sum \lambda_{vi} w_i$. Hence $M(\sum w_i)$ will have only positive coefficients for sufficiently large M, and thus will lie in $\bigcup_{m\in N} m\operatorname{Log} P$. Doing this for all the v's in Log f, and taking the maximum of the corresponding M's, yields the desired w.

Obviously f may be replaced by $x^{-w}f$, and so we may assume $\operatorname{Log} f \subset \operatorname{Log} P$. We thus have that $f/P^m$ belongs to $R_P$. Now $\operatorname{Log} Q \subset \operatorname{Log} P$ implies that Q/P lies in $R_P$. Set $u = (Q/P)^n$ (where n is such that $Q^n f$ has only positive coefficients). By [H1;V.5], $Q/P^n$ is an order unit in $R_P$, and thus so is u (see the argument in the last line in the proof of II.5 above). Obviously, $u-(f/P^m)$ is positive in $R_P$ (as $Q^n f$ has only positive coefficients), so by I.4, $f/P^m$ is positive in $R_P$. From the way the ordering on $S_P$ was defined, there exists an integer N such that $P^N f$ has only non-negative coefficients.

Now for the general case (c.f., [H2; §II]). To simplify terminology, we say (for g a polynomial) "$g \geq 0$" if g has only non-negative coefficients. Consider the inclusion of abelian groups, $Z \subset Y+Z$. Let $\{e_\theta\}_{\theta\in H}$ be a transversal (that is, Y+Z is a disjoint union of the cosets $\{e_\theta+Z\}$). For θ in the index set H, define $p_\theta: Y+Z \longrightarrow e_\theta+Z$ to be the obvious "mapping":

$$
\begin{array}{ll}
p_\theta(w) \text{ is undefined} & \text{if } w \notin e_\theta+Z \\
p_\theta(w) = w & \text{if } w \text{ belongs to } e_\theta+Z.
\end{array}
$$

Then $p_\theta$ can be defined on polynomials via

$$
p_\theta\left(\sum \lambda_w x^w\right) = \sum \lambda_w x^{p_\theta(w)},
$$

where $x^{(\text{undefined})} = 0$. It is immediate that for any polynomial g, $g \geq 0$ if and only if $p_\theta(g) \geq 0$ for all θ.

From $Q^n f \geq 0$, for all θ in H, $p_\theta(Q^n f) \geq 0$; since $\operatorname{Log} Q^n \subset Z$, this yields $Q^n p_\theta(f) \geq 0$ for all θ. Since $\operatorname{Log}(x^{-e_\theta} \cdot p_\theta(f)) \subset Z$, there exists N(θ) such that $P^{N(\theta)} p_\theta(f) \geq 0$ (the monomial with exponent $-e_\theta$ can obviously be ignored), by the result of the case $Y \subset Z$. However, for only finitely many θ is $p_\theta(f)$ nonzero. Setting N to be the maximum of the N(θ)'s for which $p_\theta(f)$ is not zero, we have that $P^N p_\theta(f) \geq 0$ for all θ. Since $\operatorname{Log} P^n \subset Z$, $P^N p_\theta(f) = p_\theta(P^N f)$, so $P^N f \geq 0$, as desired.∎

There is an obvious attempt to prove II.1; this try has a large whole in it. Namely, $\operatorname{Log} Q \subset \operatorname{Log} P$ implies that there exists M so that $Q \leq MP$ (using the notation of the last part of the proof of II.1). From $Q^n f \geq 0$, we 'deduce' $M^n P^n f \geq Q^n f \geq 0$. This is fallacious as f is not known to be positive. Moreover, the argument does not make use of the inclusion $d_e \operatorname{Log} P \subset$

Log Q, which is absolutely crucial. As referred to in the second paragraph of this chapter, there exists an f such that $(1+x+y)f$ is positive, yet $(1+x+y+xy)^N f$ is not positive for any N (see below). Finally, as in the example in the introduction, there is no obvious relation between n and N, as would be implied by the fallacious argument of this paragraph.

In one variable, with Q solid (no gaps in the monomials), the theorem is a simple consequence of an 1911 result of Meißner, which has as a consequence that $Q^n f$ has only positive coefficients for some n if and only if $f|(0,\infty) > 0$ as a function. However if Q were say $x^3+x+1$, this criterion is not now sufficient, because of the missing $x^2$ term.

Since P can obviously be replaced by $P^m$ for any m, we deduce as a trivial consequence of I.1, that if $d_e\text{Log } P^m \subset \text{Log } Q \subset \text{Log } P^m$, then the same conclusion holds. On the other hand, certain shapes of Log sets can be completely distinguished from each other via suitable f's, as in the case of the triangle and square mentioned earlier. There exists a situation in which there is only a one-sided distinction possible:

**EXAMPLE** Let $P_o$ be any solid polynomial with non-negative coefficients, let $P_1$ be another, and set $P_2 = P_1P_0$. If P, Q are polynomials with only non-negative coefficients and Log P = Log $P_o$, Log Q = Log $P_2$, then $P^n f \geq 0$; will imply $Q^N f \geq 0$ for some N ($P_o^m f \geq 0$ for some m by I.1, so $P_2^m f = P_o^m P_1^m f \geq 0$, whence $Q^N f \geq 0$ for some N by another application of I.I). It need not be true that $Q^n f \geq 0$ implies $P^N f \geq 0$. This occurs with $P_o = P = 1+x+y$, $P_1 = 1+x$, $P_2 = Q$, $f = x^2-x+1$. Then $P_1 f = x^3+1 \geq 0$, so $Qf \geq 0$; however, the coefficient of $y^N x$ in $(1+x+y)^N(x^2-x+1) = P^N f$ is $-1$, so $P^N f$ has only positive coefficients for no value of N.

This example also provides one half of the distinction between the square and the triangle (see the comment after the statement of Theorem II.1 in the introduction to this section). The other direction is given as follows.

Set $P = 1+x+y$, $Q = 1+x+y+xy = (1+x)(1+y)$, $f = x^2+y^2-xy+1$. By [H1;V.7] (or direct computation), there exists an n with $P^n f \geq 0$. On the other hand, the coefficient of $x^{N+1}y^{N+1}$ in $Q^N f = (1+x)^2(1+y)^2(x^2+y^2-xy+1)$ is $-1$, so $Q^N f$ never has only non-negative coefficients.∎

We conclude from the theorem of this section that just as $R_S$ is obtainable as a ring from $R_P$ (for any P with a Log P = b S, for some integers a, b) by inverting the order units, it is also obtainable as an ***ordered*** ring by inverting the order units; if u in $R_P$ is positive in $R_S$, it is already positive in $R_P$.

From now on, we usually (but not always) restrict $S$ to be the set of all lattice points in an integral polytope $K$; the corresponding ordered ring is denoted $R_K$.

We now consider the pure states of $R_K$. It will follow that every real algebra homomorphism $R_K \longrightarrow \mathbf{C}$ is actually a pure state (and in particular has values in $\mathbf{R}$), so that the maximal ideal space of the complexification $R_K \otimes \mathbf{C}$ is just the pure state space of $R_P$.

Recall from section I that the pure states of a partially ordered ring $R$ having 1 as an order unit are precisely the positive group homomorphisms $R \longrightarrow \mathbf{R}$ that are ring homomorphisms as well. In case $R$ is a rational or real algebra, it is easy to verify that the pure states are also algebra homomorphisms.

For $R_P$, the pure states were determined in [H1; section III]; integer coefficients were used there; the results are the same over the reals. The pure states $\alpha$ that are faithful, that is, $\ker \alpha \cap R_P^+ = \{0\}$, are given precisely as evaluations at points of $(\mathbf{R}^d)^{++}$. This means that all of the faithful pure states are constructed from a point $y = (r_1, \ldots, r_d)$ with each of the $r_i$ strictly positive real numbers; we consider elements of $R_P$ as rational functions whose denominators (powers of $P$) do not vanish at $y$, and so define the point evaluation at $y$ given by $\alpha_y(s) = s(y)$, for $s$ in $R_P$.

More generally, let $X(t) = (A_1 t^{u(1)}, \ldots, A_d t^{u(d)})$ be a path in $(\mathbf{R}^d)^{++}$, where $u = (u(1), \ldots, u(d))$ belongs to $\mathbf{Z}^d$, and $A = (A_1, \ldots, A_d)$ belongs to $(\mathbf{R}^d)^{++}$. As elements of $R_P$ are bounded as functions on $(\mathbf{R}^d)^{++}$, for $s$ in $R_P$, $s(X(t))$ will be a bounded rational function of $t$, so that $\alpha_x(s) = \lim_{t\to\infty} s(X(t))$ exists. All the non-faithful pure states of $R_P$ are of this form, and moreover only very specific $u$ and $A$ (depending only on cvx Log P) are required (see [H1; section III] for more details). In particular, $d_eS(R_P,1)$ (the set of pure states of $R_P$ with respect to 1, equipped with the point-open topology, that is, the weak topology) is a compactification of $(\mathbf{R}^d)^{++}$, and is thus connected. In section IV, we shall show that it is homeomorphic to $K$ itself, so is even contractible.

Now we consider the pure states of $R_K$, or slightly more generally, those of $R_S$, for $S =$ Log P. By II.1 and [GH1; 4.1], the inclusion map $R_P \longrightarrow R_S$ induces, by restriction, a map on the state spaces $S(R_S,1) \longrightarrow S(R_P,1)$ which is *onto* ; it is easy to check that since both of these are Bauer simplices and the map comes from a ring homomorphism, there is a corresponding onto, continuous map $d_eS(R_S,1) \longrightarrow d_eS(R_P,1)$ between compact spaces. That the latter map is one to one follows from $R_S$ being a localization of $R_P$ (througout, localization means localization at a multiplicatively closed set unless otherwise specified) and all the pure states being multiplicative. Hence the restriction map is a homeomorphism. Similar arguments yield that the inclusion maps

$R_P \subset R_S \subset R_K$ (where $K = \text{cvx Log } P$), induce homeomorphisms on the pure state spaces (by restriction). In any case we identify the pure state spaces.

Inasmuch as $R_P$, $R_S$, and $R_K$ are all unperforated, the order units are precisely the elements that are strictly positive at all pure states; alternately (from the density of the point evaluation states), bounded below on $(\mathbf{R}^d)^{++}$.

II.6 **PROPOSITION** (a) Every algebra homomorphism $R_S \longrightarrow \mathbf{C}$ is a pure state; in particular, the maximal ideals are real.

(b) If $s$ in $R_S$ does not vanish at any point of $d_eS(R_S,1)$, then either $s$ or $-s$ is an order unit, and thus $s$ is invertible.

Proof: We prove the following more general result. Let $R$ be a real algebra, all of whose field images are at most 2-dimensional over $\mathbf{R}$ (i.e., only $\mathbf{R}$ and $\mathbf{C}$ arise as quotients by maximal ideals). Let $X \subset \text{Max Spec } R$ be a compact set of maximal ideals (with respect to the point-open topology), all of whose quotients are real. Set $U = \{u \in R \mid x(u) > 0 \text{ for all } x \text{ in } X\}$. Then

$$X = \text{Max Spec } U^{-1}R.$$

Select a homomorphism $\beta: U^{-1}R \longrightarrow \mathbf{C}$, and suppose the restriction of $\beta$ to $R$ does not belong to $X$. For each $x:R \longrightarrow \mathbf{R}$ in $X$, there exists $s$ in $R$ such that $x(s) \neq 0$ and $\beta(s) = 0$ (since $\ker x$ is a maximal ideal). There exists a neighbourhood of $x$, call it $W_x$, such that for all $z$ in $W_x$, $z(s) \neq 0$. The set $\{W_x\}$ is an open covering of $X$, so by compactness, we may find a finite collection of open sets $\{W_i\}$ and corresponding $\{s_i\}$, $\{x_i \in W_i\}$ with $X \subset \cup W_i$, and $\beta(s_i) = 0$, and $z(s_i) \neq 0$ for all $z$ in $W_i$. Set $u = \sum s_i^2$. Then for all $x$ in $X$, $x(u) = \sum x(s_i)^2 > 0$, so $u^{-1}$ belongs to $U^{-1}R$. On the other hand, $\beta(u) = 0$, so $\beta(u^{-1})$ is not defined, a contradiction.

(b). For $s$ in $R_S$, we obtain a continuous function $s^\wedge: d_eS(R_S,1) \longrightarrow \mathbf{R}$; as $d_eS(R_S,1)$ is connected, the sign of $s^\wedge$ is constant. Hence one of $s$ or $-s$ is strictly positive at all pure states, so by [H1;I.1], one of them is an order unit in $R_S$ and is thus invertible therein.∎

II.7 **PROPOSITION** The maps $\{\text{ideals in } R_P\} \longrightarrow \{\text{ideals in } R_S\}$

$$\{\text{ideals in } R_S\} \longrightarrow \{\text{ideals in } R_P\}$$

given respectively by $I \mapsto IR_S$, $J \mapsto J \cap R_P$, induce inverse bijections on the sets of *order* -ideals.

Proof: If $I$ is an order ideal of $R_P$, then everything in $IR_S$ can be written in the form $iu^{-1}$ for $i$ in $I$, and $u$ an order unit of $R_P$. Suppose $0 \leq av^{-1} \leq iu^{-1}$ with $a, v$ in $R_P$ and $v$ an order unit. Multiplying by $uv$, we obtain $0 \leq au \leq iv$ in $R_S$ and thus in $R_P$ by the main result of this section. Hence $au$ belongs to $I$, so $av^{-1} = (au)(uv)^{-1}$ belongs to $IR_S$. Thus $IR_S$ is convex, and it follows from $I^+ = I^+ - I^+$ that $IR_S$ is also directed, so that $IR_S$ is an order ideal in $R_S$.

Now let $J$ be an order ideal of $R_S$. If $0 \leq a \leq b$ in $R_P$ and $b$ belongs to $J$, then as $J$ is convex, $a$ belongs to $R_P \cap J$, so that $R_P \cap J$ is also convex. As $R_S$ is noetherian, $J$ is a finitely generated ideal, so possesses a (relative) order unit (i.e., $J$ contains an element which is an order unit for $J$ considered as partially ordered abelian group) $j = ru^{-1}$. As $j$ is positive, so is $r$, and clearly $r \in R_P \cap J$. If $a$ belongs to $R_P \cap J$, as $u^{-1} \leq M \cdot 1$, for some positive integer $M$, there exists $N$ so that $a \leq Nj \leq NMr$ in $R_S$, hence in $R_P$. Thus $r$ is a relative order unit for $R_P \cap J$; thus the latter is an order ideal of $R_P$.

Clearly $(R_P \cap J)R_S \subset J$ for $J$ an order ideal of $R_S$. Select $s$ in $J^+$; write $s = au^{-1}$, with $a$ in $R_P$, $u$ an order unit of $R_P$; necessarily $a$ belongs to $R_P^+$. Then $su \in (J \cap R_P)^+$, so that $s = (su)u^{-1} \in (J \cap R_P)R_S$, and thus $J^+ \subset (J \cap R_P)R_S$; as $J$ is directed, $J \subset (J \cap R_P)R_S$, so equality holds.

Conversely, let $I$ be an order ideal of $R_P$. Obviously, $I = IR_S \cap R_P$, and the latter is an order ideal of $R_P$. If $i$ is an order unit for $I$, then it is easy to verify that it is also for $IR_S$ (within $R_S$) and thus for $IR_S \cap R_P$, so that the two order ideals must be equal.∎

Of course the point of this last result is to enable to us to pass freely between the order ideals of $R_P$ and those of the localization $R_S$.

## III INTEGRAL CLOSURE AND COHEN-MACAULEYNESS

Here we determine some sufficient conditions under which rings of the form $R_P$ or $R_K$ will be integrally closed. We show, for example, that if $K = eK'$ for some integer $e \geq d = \dim K$ and K' is an integral polytope, then $R_K$ is integrally closed (III.6), and so is any $R_P$ for P in $\mathbf{R}[x_i^{\pm 1}]^+$ with $\operatorname{Log} P = K \cap \mathbf{Z}^d$. Alternately, it is sufficient that $d(K \cap \mathbf{Z}^d) = dK \cap \mathbf{Z}^d$. Under the same hypotheses, both $R_P$ and $R_K$ are Cohen Macauley (III.2; this possibility was suggested to me by Mel Hochster); on the other hand, the general $R_P$ need not be (as Lex Renner pointed out to me).

When $R_K$ is *not* integrally closed (which forces $d \geq 3$), then its integral closure can still be computed quite easily, as a ring of bounded rational functions with suitable denominators. The integral closure need not be of the form $R_{dK}$ (which is always integrally closed), and necessary and sufficient conditions on K (more precisely, on the boundary of K) are given for this to occur (III.10). If $d = 3$, then $R_{dK}$ is automatically the integral closure of $R_K$ (for projectively faithful polytopes), but this fails for $d = 4$. In the course of proving this, an interesting result on subsemigroups of $(\mathbf{Z}^d, +)$ comes out; the result is a higher dimensional analogue of the well known fact that if $\{a_i\}$ is a collection of positive integers with 1 as greatest common divisor, then there exists an integer N so that every integer n exceeding N can be expressed as a positive integer combination of the $a_i$'s.

A portion of our discussion on the integral closure of $R_K$'s and similar rings will revolve around a straightforward reduction to localizations of monomial algebras considered by Hochster in [Ho]. Let P be an element of $\mathbf{R}[x_i^{\pm 1}]^+$, and set $K = \operatorname{cvx} \operatorname{Log} P$. Let v be a vertex of the integral polytope K; translate K so that $v = \mathbf{0}$; by applying a suitable element of $GL(d,\mathbf{Z})$, we may even assume that $K \subset (\mathbf{R}^d)^+$ as well. Let M(K) be the multiplicative semigroup generated by $\{x^w \mid w \in K \cap \mathbf{Z}^d\}$, and let $W(K) = \operatorname{Log} M(K)$ be the corresponding additive subsemigroup of $\mathbf{Z}^d$ generated by $K \cap \mathbf{Z}^d$. Form $\mathbf{R}[M(K)]$, the semigroup ring of M(K), i.e.. $\mathbf{R}[x^w \mid w \in K \cap \mathbf{Z}^d]$. This type of ring is the principal object of study in [Ho] (with $\mathbf{R}$ replaced by an arbitrary field and the semigroup W(K) having some additional hypotheses imposed upon it). Similarly, define M(P) and W(P) by replacing $K \cap \mathbf{Z}^d$ by Log P.

III.1 **LEMMA** Every localization of $R_P$ at a prime ideal is a localization of $\mathbf{R}[M(P)]$ (for some choice of vertex v).

Proof: Let J be an ideal of $R_P$ and suppose that we localize at J to form $(R_P)_J$. Set K = cvx Log P; for v in $d_eK$, set $a_v = x^v/P$ in $R_P$. The ideal generated by all of the $a_v$'s (as v varies over the vertices of K) is an order-ideal by II.2A and II.6. If it were proper, it would be contained in a maximal order ideal. The maximal order-ideals are of the form $I_v = \{x^w/P \mid w \in \text{Log } P\setminus\{v\}\}$ for some *vertex* v of K ([H1; §VII]), so that the ideal generated by the $a_v$'s is improper. (An alternate argument will be given in the proof of the next proposition.) Hence there exists a vertex v such that $a_v$ does not belong to J; hence $a_v^{-1} = Px^{-v}$ belongs to $(R_p)_J$. Now make the translation and unimodular **Z**-linear transformation so that this v is sent to the origin, and K is sent into $(\mathbf{R}^d)^+$.

Then $a_v^{-1} = P$, and so $R_P[a_v^{-1}]$ is just $R_P[P]$, which is $\mathbf{R}[M(P),P^{-1}]$, and thus is a localization of $\mathbf{R}[M(P)]$. As $(R_P)_J$ is a localization of $R_P[(a_v)^{-1}]$, we are done.∎

Thus many local properties can be obtained via results in [Ho].

Now we define a hierarchy of concepts, which measure the holes (or their lack) in multiples of $K \cap \mathbf{Z}^d$. Let K be an integral polytope, inside $\mathbf{R}^d$. We say K is <u>k-convex</u> (with k a positive integer) if

$$K \cap \mathbf{Z}^d = k(K \cap \mathbf{Z}^d).$$

The integral polytope K is <u>solid</u> if it is k-convex for all positive integers k. We shall see later that if dim K = d, and K is d-convex, then K is solid. Also, if d = 2, then K is automatically solid. In [H3], it is shown that if d = 3 and K is 2-convex, then K is solid, but I do not know if for d exceeding 3, that d–1-convexity is sufficient to guarantee solidity. On the other hand, there are examples to show that even for projectively faithful polytopes, for each d ≥ 3, d–2-convexity is not sufficient to guarantee solidity (Example III.3 may be generalized to all higher dimensions).

There are local (geometric) properties to go along with these global ones. Suppose that K is an integral polytope and v is a vertex of K. Then K is <u>locally solid at v</u> if

$$\bigcup_{k\in\mathbf{Z}} k\Big((K-v)\cap\mathbf{Z}^d\Big) = \Big(\bigcup_{k\in\mathbf{Z}} k(K-v)\Big)\cap\mathbf{Z}^d;$$

in other words, after translating so that v is the origin, any lattice point in the $\mathbf{R}^+$-cone generated by K (with origin v) is in the semigroup generated by $(K \cap \mathbf{Z}^d) - v$. We say K is <u>locally solid</u> if it is locally solid at v for every vertex v. Being solid is at least formally stronger than being locally solid.

Finally, K is integrally simple at v if the cone in $\mathbf{Z}^d$ generated by $(K\cap \mathbf{Z}^d) - v$ (in $\mathbf{Z}^d$) is simplicial; equivalently, its fundamental polytope is AGL(d,Z)-equivalent to the standard unit simplex. If K is integrally simple at each vertex v, then we call K integrally simple. This is the strongest property of all those introduced here, and we shall show (section VI) that if K is projectively faithful, then $R_K$ is factorial if and only if K is integrally simple; this obviously entails regularity of $R_K$.

**III.2 PROPOSITION** Let $P = \sum \lambda_w x^w$ be an element of $\mathbf{R}[x_i^{\pm 1}]^+$, and suppose that $\operatorname{Log} P = K\cap \mathbf{Z}^d$ where K is an integral polytope. If K is locally solid, then

(i) $R_P$ is integrally closed in its field of fractions, and

(ii) $R_P$ is Cohen Macauley.

Proof: Let J be a prime ideal of $R_P$, and consider the localization of $R_P$ at J, $(R_P)_J$. Set $b_w = x^w/P$ for w in Log P. As $\sum \lambda_w b_w = 1$, at least one of the $b_w$'s does not belong to J. Consider $a_v = x^v/P$ for v in $d_eK$. By [H1; III.3], given w in Log P, there exist positive integers n(v) adding up to N so that $b_w^N = \prod a_v^{n(v)}$. If every $a_v$ belonged to J, as J is prime, then every $b_w$ would also belong. Hence there exists a vertex v of K such that $a_v$ does not belong to J. Thus $(R_P)_J$ is a localization of $R_P[a_v^{-1}]$, which as we saw in the previous argument, is a localization of $\mathbf{R}[M(P)]$ for the vertex v. As being locally solid entails normality of each M(P) (one for each vertex) in the sense of [Ho; p. 320, P. 1], [Ho; Proposition 1 and Theorem 1] apply.∎

Even if K is projectively faithful, then $R_K$ need not be integrally closed (let alone $R_P$). The simplest and first such example is probably the following, due to Hochster and Stanley (verbal communication: I had asserted that $R_K$ was always integrally closed and they promptly disabused me of that notion with the following example):

**III.3 EXAMPLE** Let K be the convex hull of

$$S = \{(0,0,0), (1,1,0), (1,0,1), (0,1,1), (2,1,0)\}.$$

Then $K\cap \mathbf{Z}^d = S$, and S generates all of $\mathbf{Z}^d$ as an abelian group; so K is projectively faithful. On one hand, (1,1,1) belongs to 2K, but on the other, it does not not belong to nS for any n. Thus K is not k-convex for any k exceeding one.

Let $P = \sum_S x^w$; we now prove that $xyz/P^2$ cannot belong to $R_K$, although it does belong to its integral closure. If $xyz/P^2$ belonged to $R_K$, it would be positive, as it is positive in

$R[x^{\pm 1}, y^{\pm 1}, z^{\pm 1}, P^{-1}]$ with respect to the limit ordering. Hence there would exist f, Q in $\mathbf{R}[x,y,z]$, such that $\operatorname{Log} f \subset \operatorname{Log} P^k$ for some k and $Q/P^n$ is an order unit in $R_K$ so that

$$\frac{xyz}{P^2} = \frac{f}{P^k} \cdot \left(\frac{Q}{P^n}\right)^{-1};$$

as $f/P^k$ belongs to $R_K^+$ by II.1, and belongs to $R_P$, again by II.1, it belongs to $R_P^+$. Hence we may assume (multiplying numerator and denominator by a power of P) that all coefficients of f are non-negative. Moreover, as $Q/P^n$ is an order unit, we must have that $0 \in \operatorname{Log} Q$, and we may assume that all the coefficients of Q are non-negative. We have

$$xyzP^kQ = P^2 fP^n$$

with all the polynomials having only non-negative coefficients. Hence

$$(1,1,1) + k \operatorname{Log} P + \operatorname{Log} Q = (n+2) \operatorname{Log} P + \operatorname{Log} f + (n+k+2) \operatorname{Log} P.$$

As $0 = (0,0,0)$ belongs to $k\operatorname{Log} P$ and to $\operatorname{Log} Q$, this would yield $(1,1,1) \in (n+k+2) \operatorname{Log} P = (n+k+2) S$, a contradiction

To see that $xyz/P^2$ is integral over $R_P$ (hence over $R_K$), we note that $(1,1,1)$ is the barycentre of $3\operatorname{cvx}\{S \setminus (2,1,0)\}$, and it follows that

$$\left(\frac{xyz}{P^2}\right)^2 = \frac{1}{P} \cdot \frac{xy}{P} \cdot \frac{yz}{P} \cdot \frac{xz}{P}. \blacksquare$$

This last calculation generalizes quite easily, so that the integral closure may always be computed:

III.4 **PROPOSITION** Let $P = \sum \lambda_w x^w$ be a Laurent polynomial in d variables with positive coefficients. If P is projectively faithful, then the integral closure of $R_P$ in its field of fractions $\mathbf{R}(x_i)$ is:

$$\underline{R_P} = \{f/P^n \mid f \in \mathbf{R}[x_i^{\pm 1}] \text{ and } \operatorname{Log} f \subset n \operatorname{cvx} \operatorname{Log} P\}.$$

and this is the same as

$$\{\beta = f/P^n \mid |\beta(r)| \leq N \text{ for some } N \text{ in } \mathbf{N}, \text{ for all } r = (r_1, r_2, \ldots r_d) \text{ in } (\mathbf{R}^d)^{++}\}.$$

Proof: Let $\beta = f/P^n$ ($f$ in $\mathbf{R}[x_i^{\pm 1}]$) be integral over $R_P$. The equation $\beta^m + \sum s_i \beta^i = 0$ ($s_i$ in $R_P$) evaluated at any r in $(\mathbf{R}^d)^{++}$ yields $|\beta(r)| \leq \max\{(|s_i(r)|/m)^{m-i}\}$; as the $s_i$ belong to $R_P$, these are bounded as functions on $(\mathbf{R}^d)^{++}$, so $\beta$ is also bounded.

Now suppose simply that $\beta$ is bounded (as a function on $(\mathbf{R}^d)^{++}$). We shall show that $\mathrm{Log}\, f \subset n\,\mathrm{cvx}\ \mathrm{Log}\ P$. If not, $L = \mathrm{cvx}\{\mathrm{Log}\, f \cup n\,\mathrm{Log}\, P\}$ has a vertex $v$ in $\mathrm{Log}\, f$ but not in $\mathrm{cvx}(n\,\mathrm{Log}\, P)$. Let $u$ in $\mathbf{R}^d$ be a linear functional exposing $v$, i.e., $u \cdot v \geq u \cdot w$ for all $w$ in $L \setminus \{v\}$. Define the path $X(t) = (\ldots, t^{u(i)}, \ldots)$ for $t \geq 0$ (here $u = (u(1), u(2), \ldots, u(d))$, so this is a path in $(\mathbf{R}^d)^{++}$. It is helpful to think of this as the exponential of a straight line through the origin in $\mathbf{R}^d$. Let $\alpha$ be the coefficient of $x^v$ in f; then $\alpha$ is not zero. We have that $\beta(X(t)) = (\alpha t^{u\cdot v}$ + terms of lower degree) divided by (terms of lower degree). Recalling that the coefficients of the denominator are positive, for large t, $\beta(X(t))/t^{u\cdot v} \sim \alpha$, so that $\beta$ is unbounded along this path, a contradiction. Thus $\mathrm{Log}\, f \subset n\,\mathrm{cvx}\,\mathrm{Log}\, P$ (c.f., [H1:§III]).

Next we show that if $w \in \mathbf{Z}^d \cap n\,\mathrm{cvx}\mathrm{Log}\ P$ then $x^w/P^n$ is integral over $R_P$. We may write $w = \sum c_v\, n v$ with $c_v$ rational, non-negative and the v's running over the vertices of $K = \mathrm{cvx}\,\mathrm{Log}\ P$, and with $\sum c_v = 1$ [H1;III.1A]. There exists a positive integer N so that all of the terms $c_v n N$ are integers. As $d_e K = \mathrm{Log}\ P$, we have that

$$\left(\frac{x^w}{P^n}\right)^N = \prod_{v \in d_e K} \left(\frac{x^v}{P}\right)^{c_v n N}$$

expresses $x^w/P^n$ as an N-th root of an element of $R_P$. This yields that $\underline{R}_P$ is contained in the integral closure, and the preceding arguments yield all the desired equalities.∎

It will follow from our d-convex results below (III.7) that $\underline{R}_P$ is always generated as a ring by $\{x^w/P^d \mid w \in \mathrm{cvx}(d\mathrm{Log}\ P) \cap \mathbf{Z}^d\}$, so is finitely generated as a ring, and as an $R_P$-module. However, the integral closure of $R_K$ is not generally $R_{dK}$ (see the comment just before Lemma III.9, together with III.10).

III.5 **LEMMA** Let K be an integral polytope of dimension d inside $\mathbf{R}^d$. Suppose $K = \cup K_i$ where each $K_i$ is also an integral polytope. If each $K_i$ is k-convex for some fixed k, then so is K. In particular, if the $K_i$'s are all solid, then K is solid.

Proof: Assume that all the $K_i$'s are k-convex. Select x in $kK \cap \mathbf{Z}^d$. Then $x/k$ belongs to $K_i$ for some i, so that x belongs to $kK_i$. As $K_i$ is k-convex, we may find $v_1, \ldots, v_k$ in $K_i \cap \mathbf{Z}^d$ (hence in $K \cap \mathbf{Z}^d$) so that $x = \sum v_i$. Thus $kK \cap \mathbf{Z}^d \subset k(K \cap \mathbf{Z}^d)$. The reverse inequality is trivially true.∎

**III.6 LEMMA** Let K be an integral polytope in $\mathbf{R}^d$ which is a simplex and has interior. Then for all $e \geq d$, $eK$ is solid.

Proof: Fix k; we show that $eK$ is k-convex. By translating if necessary, we may assume that the set of vertices of K is the affinely independent set $\{\mathbf{0}, v_1, \ldots, v_d\} \subset \mathbf{Z}^d$. Select x in $keK$. We will find $x^1$ in $eK \cap \mathbf{Z}^d$, and $x^2$ in $(k-1)eK \cap \mathbf{Z}^d$ so that $x = x^1 + x^2$. Then the result follows by induction on k (eK is of course 1-convex in any case).

There exist affine linear functionals $d_i$, $i = 1,2,\ldots,d$, together with $D = 1 - \sum d_i$ such that $d_i(v_j) = 1$ if $i = j$, 0 else, and so that

$$K = \cap d_i^{-1}([0,\infty)) \cap D^{-1}([0,\infty)).$$

Now $keK$ has vertices $\{0, kev_i\}$, so we have that $ke \geq d_i(x) \geq 0$ and $\sum d_i(x) \leq ke$. We apply the following process to x.

If $d_1(x) \geq 1$, find the integer $b'_1$ so that $b'_1 \leq d_1(x) < b'_1 + 1$; set $b_1 = \min\{e, b'_1\}$. Assume $b_1, b_2, \ldots, b_r$ have been defined so that $\sum_{i \leq r} b_i < e$ (if at any step, the sum equals e, stop), and $b_i \leq d_i(x) < b_i + 1$, for $i \leq r$, the $b_i$ all being integers. Define $b_{r+1}$ as $\min\{e - \sum_{i \leq r} b_i, b'_{r+1}\}$ where $b'_{r+1} = [d_{r+1}(x)]$.

If after s steps in this process, $\sum_{i \leq s} b_i = e$, set $x^1 = \sum_{i \leq s} b_i v_i \in \mathbf{Z}^d$. We see that $d_j(x_1) = b_j \leq e$ if $j > s$, and $d_j(x^1) = 0$ if $j \leq s$, and moreover, $(\sum d_i)(x^1) \leq \sum b_i = e$, so that $x^1$ belongs to $eK \cap \mathbf{Z}^d$. Furthermore, $x^2 = x - x_1$ satisfies $d_i(x^2) \geq 0$ and $(\sum d_i)(x^2) = \sum d_i(x) - e \leq (k-1)e$, so $x^2$ belongs to $(k-1)eK \cap \mathbf{Z}^d$.

If the process described above does not terminate, we end up with $x^1 = \sum_{i \leq d} b_i v_i$, with $\sum_{i \leq s} b_i \leq e$. By construction, for each i, $d_i(x) - b_i < 1$, so if $x^2 = x - x^1$, $d_i(x^2) < 1$, whence $\sum d_i(x^2) \leq d \leq e$. Therefore $x^2$ belongs to $eK \cap \mathbf{Z}^d$ ($x^1$ belongs to $\mathbf{Z}^d$, and thus so does $x^2$). This latter set is contained in $(k-1)eK \cap \mathbf{Z}^d$ as $\mathbf{0}$ belongs to K. Obviously $x^2$ belongs to eK (as $\sum b_i \leq e$) and we are done.∎

**III.7 PROPOSITION** If K is a d-dimensional integral polytope in $\mathbf{R}^d$, then for all $e \geq d$, eK is solid. In particular, $R_{dK}$ is integrally closed and $R_{dK} = R_{eK}$ for all $e \geq d$.

Proof: Every such K can be triangulated by simplices $K_i$ which are integral polytopes. Then the previous two results apply. The second statement comes from III.1 and the definition of solid.∎

**REMARK:** It is probably true that the $e \geq d$ conditions in III.6, III.7 can be replaced by $e \geq d-1$. This will be seen below if $d = 2$, and is proved in the case that $d = 3$ in [H3;Corollary, Section VI]. If K happens to be a right simplex (or can be integrally triangulated by such), meaning that its vertices are $\{\mathbf{0}, (a_1,0,\dots,0), (0,a_2,0,\dots,0),\dots, (0,\dots,0,a_d)\}$, then in this case as well $e \geq d$ may be replaced by $e \geq d-1$. On the other hand, by adapting III.3 appropriately, one can construct, for every value of $d = \dim K$ exceeding 2, integral polytopes K for which $(d-2)K$ is not 2-convex.

If K is the 3-dimensional right simplex with $a_1 = 5$, $a_2 = 3$, and $a_3 = 2$, then K is projectively faithful, but as Ed Formanek pointed out, it is not locally solid.

III.8 **LEMMA** Let T be an integral triangle in $\mathbf{R}^2$. If $T \cap \mathbf{Z}^d = d_e T$, then T is AGL(2,Z)-equivalent to the standard unit triangle, that with vertices $\{(0,0), (1,0), (0,1)\}$.

Proof: By translation, one of the vertices of T may be assumed to be (0,0); by a unimodular change of variables, the second vertex can be assumed to be (1,0) (since the greatest common divisor of the coordinates must be one); call the third vertex (r,s). If $r = 0$ and $s = 1$, then we are done. If $s < 0$, the transformation $(a,b) \mapsto (a,-b)$ is in GL(2,Z), and so allows us to assume $s > 0$. Matrix multiplication by $\begin{bmatrix} 1 & 0 \\ k & 1 \end{bmatrix}$ is admissible and changes $\{(0,0), (1,0), (r,s)\}$ to $\{(0,0), (1,0), (r+sk, s)\}$. So we may assume $s > 1$ and $0 \leq r \leq s$. Clearly the greatest common divisor of r and s must be one. Now

$$(1,1) = s^{-1}(r,s) + (1-rs^{-1})(1,0) + ((r-1)s^{-1})(0,0)$$

and the coefficients are non-negative. As s exceeds 1, we arrive at a contradiction, and so $s = 1$ and $r = 0$.∎

**COROLLARY** If K is a 2-dimensional integral polytope, then $R_K$ is integrally closed and Cohen-Macauley.

Proof: By III.5 and III.8, K is solid, so that by III.2 the desired conclusions follow.∎

III.8A **PROPOSITION** If K is a projectively faithful integral polytope, then $R_K$ being integrally closed implies $R_K = R_{2K} = \dots$, and K is locally solid.

Proof: Let $P = \sum x^w$, w varying over $K \cap \mathbf{Z}^d$ and let Q be an element of $\mathbf{R}[x_i^{\pm 1}]^+$ such that $\mathrm{cvxLog}\, Q = mK$ for some integer m. Then $s = Q/P^m$ belongs to $\underline{R}_P$ by III.4, and thus (as $R_K$ is integrally closed), $Q/P^m$ belongs to $R_K$. As $Q/P^m$ is an order unit in $R_{mK}$ (by [H1;V.4]), it is also an order unit in $R_K$ (as the pure states are the same as those of $R_{mK}$); thus $s^{-1}$ belongs to $R_K$. However, we have

$$R_Q \subset R_P[s^{-1}] \subset R_K \subset R_{mK};$$

as $R_{mK}$ is a localization of $R_Q$ obtained by inverting all of its order units, and the order units of $R_K$ are already invertible, it follows that $R_K = R_{mK}$.

Suppose that $v$ is a vertex of K; we may assume $v = \mathbf{0}$ and $K \subset (\mathbf{R}^d)^+$. Suppose $w$ belongs to $mK \cap \mathbf{Z}^d$ for some m; we wish to show that $w \in n(K \cap \mathbf{Z}^d)$ for some n, and this will show that K is locally solid. We have that $x^w/P^m \in R_K$, by the preceding paragraph, so we may write $(x^w/P^m)(f/P^t) = g/P^u$, where t,u are integers, $\operatorname{Log} g \subset u \operatorname{Log} P$, and $f/P^t$ is an order unit of $R_P$. By multiplying numerator and denominator by a power of P, we may assume that $\operatorname{Log} f \subset \operatorname{Log} P^t$, and f has only positive coefficients. Clearing denominators, we obtain $x^w f = P^{m+t-u} g$.

If $m+t \geq u$, then w belongs to $t\operatorname{Log} P + (m+t-u)\operatorname{Log} P + \operatorname{Log} g \subset (m+t)\operatorname{Log} P$; as $\mathbf{0}$ belongs to Log P, we obtain $w \in (m+t)\operatorname{Log} P = (m+t)(K \cap \mathbf{Z}^d)$. If $m < t+u$, the power of P moves to the left, and the same argument works.∎

It is entirely plausible that a locally solid integral polytope is solid, but I do not know if this is the case.

Now we determine under which conditions $R_K$ has $R_{dK}$ as its integral closure; since we are assuming K is projectively faithful, this is equivalent to deciding when $R_{dK}$ is a finitely generated $R_K$-module (the fields of fractions of $R_K$ and of $R_{dK}$ are the same). This happens to be automatic if $d \leq 3$ (in the cases that d is 1 or 2, $R_K$ is already integrally closed and equals $R_{dK}$), but not for higher d. For K a projectively faithful d-dimensional integral polytope, it follows from III.7 that $R_{dK}$ is integrally closed in its field of fractions.

Let S be any finite subset of $\mathbf{Z}^d$, and let F be a (proper) face of $L = \operatorname{cvx} S$ in $\mathbf{R}^d$. We say S is <u>projectively faithful at F</u> if the set of differences $S \cap F - S \cap F$ spans (as an abelian group) the group generated by $\bigcup_{m \in \mathbf{N}} \{(mF \cap \mathbf{Z}^d) - (mF \cap \mathbf{Z}^d)\}$.

For example, if $d = 2$, $S = \{(0,0), (2,0), (0,1), (1,1)\}$, and $F = [0,2] \times \{0\}$, then $F \cap \mathbf{Z}^2 - F \cap \mathbf{Z}^2$ spans $\mathbf{Z} \times \{0\}$, while $S \cap F - S \cap F = \{(0,0), (2,0), (-2,0)\}$ only spans $2\mathbf{Z} \times \{0\}$. We say S is <u>totally faithful</u> if (i)S is itself projectively faithful (i.e., S−S generates $\mathbf{Z}^d$ as an abelian group) and (ii)S is projectively faithful at F for each proper face F of cvx S.

Finally, the integral polytope $K$ is totally faithful if the finite set $K \cap \mathbf{Z}^d$ is so. By III.8, if $d = \dim K = 3$ and $K$ is projectively faithful, then $K$ is totally faithful.

**EXAMPLE** In four dimensions, a simple example of a projectively faithful but not totally faithful integral polytope can be constructed by making

$$\operatorname{cvx}\{(1,1,0), (1,0,1), (0,1,1), (0,0,0)\} \times \{0\}$$

a proper face of $K$ by adding some lattice points (not in the affine span of this potential face) to ensure projective faithfulness, e.g., $(0,0,0,1)$ and $(0,2,1,1)$. We shall show in III.10, for $K$ projectively faithful, that $R_{dK}$ is the integral closure of $R_K$ if and only if $K$ is totally faithful.∎

The following may be well known to semigroup theorists, but I have not been able to find any references to such a result:

III.9 **LEMMA** Let $A$ be a subsemigroup of $\mathbf{Z}^d \subset \mathbf{R}^d$ (under addition), such that $\mathbf{0} \in A$ and $A - A = \mathbf{Z}^d$. Let $b$ be an element of $\mathbf{Z}^d \cap \operatorname{Int} \operatorname{cvx} A$ (the interior computed within $R_P$). Then there exists an integer m such that for all $n \geq m$, $nb$ belongs to $A$, and moreover $m$ is independent of the choice of $b$. If for each face $F$ of $\operatorname{cvx} A$, we have that $A \cap \mathbf{Z}^d - A \cap \mathbf{Z}^d = F \cap \mathbf{Z}^d - F \cap \mathbf{Z}^d$, then $b$ belonging to $\operatorname{cvx} A \cap \mathbf{Z}^d$ is sufficient.

Proof: We first remark that if $d = 1$, this is well-known—if $\{m_1, \ldots, m_k\}$ is a set of positive integers whose greatest common divisor is one, then for all sufficiently large $n$, $n$ belongs to the semigroup generated by $\{m_i\}$. This is easy to verify. We will apply this to $\{s \in \mathbf{N} \mid sb \in A\}$.

First, if $\{0, w_1, \ldots, w_d\}$ is an affinely independent subset of $A$ (that is, $\{w_1, \ldots, w_d\}$ is $\mathbf{R}$-linearly independent), and $A'$ is the semigroup generated by $\{0, w_1, \ldots, w_d\}$, and $b$ belongs to $\mathbf{Z}^d \cap \operatorname{cvx} A'$, then $kb \in A' \subset A$, where $k$ is the absolute value of the determinant of the matrix whose columns are the $w_i$'s.

To see this, let $B = A' - A'$, so that $\mathbf{Z}^d/B$ has cardinality $k$. Then $kb$ belongs to $B$, so we may write $kb = \sum_{1 \leq j \leq d} t_j w_j$, with $t_j$ in $\mathbf{Z}$. On the other hand, there exist non-negative real numbers $r_j$ such that $b = \sum r_j w_j$ (as $b$ belongs to $\operatorname{cvx} A'$). As the set of $w_i$'s is linearly independent, we must have that $t_j = kr_j \geq 0$, so $kb$ belongs to $A'$.

By for example, Caratheodory's theorem, given $b$ satisfying the hypotheses of the lemma, there exists an $\mathbf{R}$-linearly independent subset $\{w_1, \ldots, w_d\}$ of $A$ (as $A - A = \mathbf{Z}^d$, $A \cdot \mathbf{R}^+$ is a $d$-dimensional cone) such that $b$ belongs to $\operatorname{Int} \operatorname{cvx} \{w_i\}$. If $k = \det(w_i)$, we deduce that $kb$ belongs to $A$, by the preceding paragraph.

Let p be a prime number dividing k, and set $B = \sum \mathbf{Z}w_i$. There exists a in $\mathbf{Z}^d$ so that pa belongs to B but a does not (i.e., a+B has order p in $\mathbf{Z}^d/B$). Since $A-A = \mathbf{Z}^d$ and $\mathbf{Z}^d/B$ is finite, it is easy to find such an element a in A itself. We may write $pa = \sum m_j w_j$, with $m_j$ in $\mathbf{Z}$. There exists at least one i in $\{1,2,...,d\}$ such that $(p, m_i) = 1$. Fix one choice for i.

For this choice of i, the sequence $\{w_i + a/N\}_{N\in\mathbf{N}}$ converges to $w_i$, so there exists M so that for all $N \geq M$,

$$C_N = \{w_1,...,w_{i-1}, w_i+a/N, w_{i+1},..., w_d\}$$

is linearly independent, and b belongs to Int cvx $C_N$.

Select such an N so that additionally $N-m_i$ is not divisible by p and $(N, k) = 1$ (we may even choose N to be prime!). Then b belongs to the interior of the convex hull of $\{w_1,...,w_{i-1},Nw_i+a, w_{i+1},...,w_d\}$, which is contained in Int cvx A. Now the determinant k', of the matrix with columns $\{w_1,...,w_{i-1},Nw_i+a,w_{i+1},...,w_d\}$, is

$$N\det(w_i)+\det(\{w_1,...,w_{i-1},a,w_{i+1},...,w_d\}).$$

As $a = \sum (m_j/p)w_j$, the second determinant is $\pm(m_i/p)k$, so that $k' = kN \pm m_i k/p$. Hence $(k,k') = (k, m_i k/p)$, which divides $k/p$.

Now for any prime q (including p itself) dividing k/p, we can find a similar independent subset of A whose matrix has determinant k", where (k', k") divides k'/q. Therefore, after a finite number of iterations of this process, we obtain integers $k_1 = k$, $k_2 = k'$, etc., with the greatest common divisor of $k_1, k_2,...$ equalling one. By the results above, each $k_j b$ belongs to A, and thus $\{s \in \mathbf{N} \mid sb \in A\}$ contains $\{m,m+1,m+2,...\}$ for some m, as desired.

If the facial hypotheses hold, and $b \in \text{cvx } A \setminus \text{Int cvx } A$, then b belongs to the relative interior of some face F of cvx A. Then we work inside the **R**-affine span of F (with $F\cdot\mathbf{R}$, $(A\cap F)\cdot\mathbf{R}^+$ respectively playing the rôles of $\mathbf{R}^d$, $A\cdot\mathbf{R}^+$), and the previous result applies here.∎

III.10 **THEOREM** Let K be a projectively faithful integral polytope in $\mathbf{R}^d$. Then the integral closure of $R_K$ in its field of fractions is $R_{dK}$ if and only if K is totally faithful.

<u>Proof</u>: Suppose that K is totally faithful. Set $S = K \cap \mathbf{Z}^d$ and $P = \sum_S x^w$. Form $R_P$ and its integral closure, $\underline{R}_P$, which is described in III.4. We first note that $R_P \subset R_{dK}$ and the latter is integrally closed by III.7; thus $\underline{R}_P \subset R_{dK}$. We shall show that for any element s of $\underline{R}_P$ that is an order unit (with respect to the ordering on $R_{dK}$—$\underline{R}_P$ has not been given an ordering, so it does not make sense to talk of order units with respect to it), there exists an order unit s' in $\underline{R}_P$ such that

$ss'$ belongs to $R_P$. It follows that $ss'$ is invertible in $R_K$, and this will be sufficient to prove $R_{dK}$ is integral over $R_K$.

Let $v$ be any vertex of $K$; let $A_v$ be the semigroup $\bigcup n\{(K \cap Z^d)-v\}$. As $K$ is projectively faithful, $A_v - A_v = Z^d$. By the lemma above, for all $w$ in $dK \cap Z^d$, there exists a positive integer $k = k(v)$ so that for all integers $L \geq 2^k$, we have that $L(w-dv) \in A_v$. (As $dK \cap Z^d$ is finite, we can take a single $k$ to work for all the $w$'s.)

Fix $L \geq 2^k$. There exists $m$ so that for all $w$ in $dK \cap Z^d$, $L(w-dv)$ belongs to $m\{(K \cap Z^d)-v\}$. We may assume $m \geq Ld$, as $\mathbf{0} \in K-v$. Thus

$$(1_v) \qquad \left(\frac{x^w}{P^d}\right)^L \cdot \left(\frac{x^v}{P}\right)^{m-Ld} \in R_P.$$

We may assume that the $k(v)$, and the corresponding $L, m$ are the same for all vertices $v$ of $K$, so that $(1_v)$ holds for each $v$. The ideal $I$ generated by $\{(x^v/P)^{m-Ld} \mid v \in d_eK\}$ is an order ideal of $R_P$, and as no pure state will vanish on at least one of the $x^v/P$'s (as $v \in d_eK$), we must have that $I$ is improper. Hence there exist $r_v$ in $R_P$ such that $1 = \sum r_v (x^v/P)^{m-Ld}$. Applying this to the equations in $(1_v)$ above, we obtain

$$(2) \qquad (x^w/P^d)^L \in R_P \quad \text{for all } L \geq 2^k, \text{ and all } w \text{ in } dK \cap Z^d.$$

[This does not seem to imply that w belongs to $Ld \operatorname{Log} P$.]

Index the $w$'s in $dK$, $w(1), w(2), \ldots, w(M)$. Define elements of $\underline{R}_P$ as follows:

$$a_1 = \left(\frac{x^{w_1}}{P^d}\right)^{2^{k-1}}, \quad a_2 = \left(\frac{x^{w_2}}{P^d}\right)^{2^{k-1}}, \quad \ldots, \quad a_M = \left(\frac{x^{w_M}}{P^d}\right)^{2^{k-1}};$$

$$a_{M+1} = \left(\frac{x^{w_1}}{P^d}\right)^{2^{k-2}}, \quad a_{M+2} = \left(\frac{x^{w_2}}{P^d}\right)^{2^{k-1}}, \quad \ldots, \quad a_{2M} = \left(\frac{x^{w_M}}{P^d}\right)^{2^{k-2}};$$

$$\ldots$$

$$a_{(k-1)M+1} = \frac{x^{w_1}}{P^d}, \quad \ldots \qquad \ldots \qquad \ldots, \quad a_{kM} = \frac{x^{w_M}}{P^d}.$$

Define the $R_P$-modules, $S_{-i} = S_o = R_P$, $S_i = S_{i-1}+a_i S_{i-1}$, up to $S_{kM} = \underline{R}_P$. Since $a_i^2 \in S_{i-M}$ and $S_{i-1} \subset S_i$, we see that the $S_i$'s are all *rings*.

Now let $s$ in $S_i$ be an order unit of $R_{dK}$. We show by induction that there exists $s'$ in $\underline{R}_P$ that is an order unit (with respect to $R_{dK}$) such that $ss'$ belongs to $S_{i-1}$. If $i \leq 0$, $s'$ will do. Otherwise, $i > 0$, and we may write $s = s_o - a_i s_1$ with $s_o$, $s_1$ in $S_{i-1}$.

Let $N$ be a positive integer to be determined later, and set $s' = s_o + a_i s_1 + Na_i^{2^k}$. Then $ss' = s_o^2 - a_i^2 s_i^2 + Na_i^{2^k} s_o - Na_i^{2^{k+1}} s_1$. By (2) and the fact that $a_i^2$ belongs to $S_{i-1}$, we have that $ss'$ belongs to $S_{i-1}$. We show that N can be found so that $s'$ is an order unit. Let $\beta$ be a pure state of $R_{dK}$.

If $\beta(a_i s_1) \geq 0$, then $\beta(s') \geq 0 \geq \beta(s_o) \geq \beta(s) > 0$, regardless of the value of N. As $s_1$ belongs to $R_{dK}$, there exist $q'$ in **R** and $q$ in $\mathbf{R}^+$ such that $\beta(a_i) < q'$ and $|\beta(s_1)| < q$ for all $\beta$. As $s$ is an order unit of $R_{dK}$, there exists a real $\alpha > 0$ such that $\beta(s) > \alpha$ for all $\beta$.

Set $r = \alpha/2q$, and let $N$ be an integer exceeding $(2q-\alpha)/r^{2^k}$. We have that $|\beta(as_1)| \leq \beta(a)\,|\beta(s_1)| \leq q$ (as $a = x^w/P$ and $\beta(a) \leq 1$), so that $\beta(s + 2as_1) \geq \alpha - 2q$. If $\beta(a) > r$, then $\beta(a^{2^k}) > r^{2^k}$, so $\beta(s + 2as_1 + Na^{2^k}) > 0$. On the other hand, if $\beta(a) \leq r$, then $|\beta(as_1)| \leq rq$, so $\beta(s + 2as_1) > \alpha - 2rq = 0$, and thus $\beta(s + 2as_1 + Na^{2^k}) > 0$.

As $\beta(s') > 0$ for all pure states $\beta$ of $R_{dK}$, $s'$ is an order unit thereof. This completes the induction. Hence given an order unit $s$ in $R_{dK}$, there exists an order unit $s'$ of $R_P$ with $ss'$ in $R_P$ (and of course $ss'$ is thus an order unit of $R_P$).

Now set $Q = \sum_{dK \cap \mathbf{Z}^d} x^w$, and form $R_Q$ inside $R_{dK}$. Let $s$ be an order unit of $R_Q$. We may write $s = f/Q^m$, with $\mathrm{Log}\, f \subset m(dK \cap \mathbf{Z}^d)$; as $s$ is an order unit, we have that $\{mv \mid v \in d_eK\}$ is contained in $\mathrm{Log}\, f$, and we may assume $f \in \mathbf{R}[x_i^{\pm 1}]^+$. Also, $P^d/Q$ is an order unit of $R_Q$, and $Q/P^d$ belongs to $\underline{R}_P$, and finally, $s(P^d/Q)^{-m} = f/P^{dm}$ which belongs to $\underline{R}_P$.

It follows that $R_{dK}$ is obtained by inverting all the order units of $R_P$ and adjoining their inverses to $\underline{R}_P$. Thus

$$R_{dK} = R_K[x^w/P^d \mid w \in dK \cap \mathbf{Z}^d],$$

so $R_{dK}$ is integral over $R_K$ (as each of the generators, $x^w/P^d$, $w \in dK \cap \mathbf{Z}^d$ is integral over $R_K$).

Now we prove the converse. Let $F$ be a face of $K$ with $K$ projectively faithful but not projectively faithful at $F$. Select $w$ in $(nF \cap \mathbf{Z}^d) \cup n(\mathrm{Log}\, P \cap F)$; then $x^w/P^n \in \underline{R}_P$ (since $w \in nK \cap \mathbf{Z}^d$), and $x^w/P^n$ is bounded as a function on $(\mathbf{R}^d)^{++}$. Define the real number $e = \sup\{r^w/P(r) \mid r \in (\mathbf{R}^d)^{++}\}$, and let $s = e \cdot 1 + x^w/P^n$. Obviously $s$ belongs to $\underline{R}_P$ and an order unit of $R_{dK}$. We will show that there exists no order unit of $\underline{R}_P$, $s'$, such that $ss'$ belongs to

$R_P$. Suppose $s' = f/P^m$ is such; writing $f = \sum \alpha_v x^v$ (with real $\alpha_v$), we may assume $\alpha_v \geq 0$ and of course, $\mathrm{Log}\, f \subset mK$ and $\{mv \mid v \in d_eK\} \subset \mathrm{Log}\, f$.

Now $(e \cdot 1 + x^w/P^n)(\sum \alpha_v x^v/P^m)$ belongs to $R_P$; because $F$ is a face, we have that

$$\mathrm{Log} \left\{ \left(1 + x^w\right) \left( \sum_{m \mathrm{Log}\, P \cap F} \alpha_v x^v \right) P_F^k \right\} \subset (m+n+k)(\mathrm{Log}\, P \cap F)$$

for some k (see [H1; §VII] and section I for the notation $P_F$); we may absorb $P^k$ into f (replacing $f/P^k$ by $fP^k/P^{k+m}$), so that $k = 0$. Hence $w + \{mv \mid v \in d_eK\} \subset (m+n)(\mathrm{Log}\, P \cap F)$. As $d_eF$ is contained in Log P, we would obtain that w belongs to $n'(\mathrm{Log}\, P \cap F)$ for some n', a contradiction.∎

**COROLLARY** If K is a projectively faithful integral polytope, with $\dim K \leq 3$, then $R_K$ is an order in $R_{2K}$, and the latter is integrally closed.

<u>Proof:</u> If $d \leq 2$, then $R_K = R_{2K}$, and both are integrally closed. If $d = 3$, it follows from III.6, that K is projectively faithful at all of its faces, and thus is totally faithful.∎

## IV PROJECTIVE $R_K$-MODULES ARE FREE

Over $R_K$ (as opposed to $R_P$), finitely generated projective modules are free.

To prove this, we elaborate on a special case of a topological result in the book [Ro]. Let P $= \sum \lambda_w x^w$ be in $\mathbf{R}[x_i^{\pm 1}]^+$, and set $K = \text{cvx Log } P$. In [H1; p. 50], there was defined a map between compact spaces,

$$\Lambda_P : d_e S(R_P,1) \longrightarrow K$$

$$\Lambda_P(\gamma) = \sum \lambda_w \gamma(x^w/P) w.$$

Allowing S to equal Log P, $\Lambda_P$ extends to a map, also called $\Lambda_P$, from $d_e S(R_S,1)$ to K, as the two pure state spaces may be identified with each other. We shall show that $\Lambda_P$ is always a homeomorphism. We may assume that the set of differences, S–S, generates $\mathbf{Z}^d$ as an abelian group, and thus Int K (computed in $\mathbf{R}^d$) is not empty. First, consider the effect of $\Lambda_P$ (which we now call $\Lambda$, until further notice) on the point evaluation states, the set of which we identify with $(\mathbf{R}^d)^{++}$. This yields a map,

$$\Lambda : (\mathbf{R}^d)^{++} \longrightarrow K$$

$$\Lambda(r) = \frac{\sum \lambda_w r^w w}{P(r)} .$$

Clearly $\Lambda((\mathbf{R}^d)^{++}) \subset \text{Int } K$. Writing $r = \exp(s)$ with s in $\mathbf{R}^d$, we have a corresponding map $\Gamma : \mathbf{R}^d \longrightarrow \text{Int } K$, given by

$$\Gamma(s) = \frac{\sum \lambda_w \exp(s \cdot w)\, w}{\sum \lambda_w \exp(s \cdot w)} .$$

Define an atomic measure $\mu$ on K via $\mu(\{v\}) = \lambda_v / \sum \lambda_w$ for v in $K \cap \mathbf{Z}^d$, and set $Q(s) = \int_K \exp(w \cdot s)\, d\mu(w)$. Then $\Gamma = \nabla \ln Q$, and this fits into the context of [Ro; Theorem 26.5]. (Appendix E contains a self-contained proof of this result.) Specifically, $\Gamma$ is a homeomorphism onto its image, and its image is convex. Now every vertex can be approximated by points in the image of S: If v is a vertex, let u be a supporting linear functional (that is, $u \in \mathbf{R}^d$ and $u \cdot v > u \cdot k$ for all k in K unequal to v), and consider for t in $\mathbf{R}$,

$$u\cdot\Gamma(ut) = \frac{\int_K \exp(w\cdot ut)\, u\cdot w\, d\mu(w)}{\int_K \exp(w\cdot ut)\, d\mu(w)}.$$

Simple estimates reveal that $\lim_{t\to\infty} u\cdot\Gamma(ut) = u\cdot v$, so $v$ lies in the closure of the image of $\Gamma$. As $\Gamma(\mathbf{R}^d)$ is open and convex, $\Gamma(\mathbf{R}^d) = \operatorname{Int} K$ (this is true for *any* measure, not necessarily atomic, on K, provided that the convex hull its support is all of K). Thus $\Lambda:(\mathbf{R}^d)^{++} \longrightarrow \operatorname{Int} K$ is an onto homeomorphism. This enables us to prove that $\Lambda_P$ is a homeomorphism.

Let $\gamma$ be a pure state of $R_P$. If $\gamma$ is faithful, then $\gamma$ is a point evaluation at some $r$ in $(\mathbf{R}^d)^{++}$ [H1; III.3]. If it is not faithful, $T = \{w \in \operatorname{Log} P \mid \gamma(x^w/P) \neq 0\}$ is a relative face of K, that is, $\operatorname{cvx} T = F$ is a face of K, and $F \cap \operatorname{Log} P = T$ [H1; III.2]. Clearly, $\Lambda_P$ belongs to F, and since this holds for all pure states, $\Lambda_P(\gamma)$ does not belong to any proper subface of F. The ideal $\mathfrak{p}_F$, generated by $\{x^w/P \mid w \in \operatorname{Log} P\backslash T\}$ is an order ideal and by [H1; §VII], $R_P/\mathfrak{p}_F$ is order isomorphic (as ordered rings) to $R_{P_F}$, where as usual, $P_F = \sum_{w\in T} \lambda_w x^w$. By induction on dimension, $\Lambda_{P_F}: d_eS(R_{P_F},1) \longrightarrow F$ is a homeomorphism (the 0-dimensional version is trivial to prove, so the induction can be started). Notice that the quotient map $\pi_F: R_P \longrightarrow R_P/\mathfrak{p}_F$ induces $\pi_F^*: d_eS(R_{P_F},1) \longrightarrow d_eS(R_P,1)$ (via $\pi_F^*(\gamma) = \gamma\circ\pi_F$) and $\Lambda_P\pi_F^* = \Lambda_{P_F}$. Hence each face of K lies in the image of $\Lambda_P$, and thus $\Lambda_P$ is onto.

If $\Lambda_P(\gamma) = \Lambda_P(\gamma')$ for pure states $\gamma$ and $\gamma'$ of $R_P$, then the face of $\gamma$, $F(\gamma)$, must equal that of $\gamma'$ by the preceding argument. However, we would then have that $\Lambda_{P_{F(\gamma)}} = \Lambda_{P_{F(\gamma)}} = \Lambda_F$, say, so both induce pure states $\bar\gamma, \bar\gamma'$ on $R_{P_F}$. As their images in F under $\Lambda_{P_F}$ are equal, and by induction $\Lambda_{P_F}$ is one to one, it follows that $\bar\gamma = \bar\gamma'$ so $\gamma = \gamma'$. Thus $\Lambda_P$ is one to one and onto. Clearly $\Lambda_P$ is continuous, and since both $d_eS(R_P,1)$ and K are compact, $\Lambda_P$ is a homeomorphism.

In view of the description of the pure states of any $R_Q$ given in [H1; §III], it is clear that every pure state of $R_P$ extends uniquely to a pure state of $R_K$, and this identification of the pure state spaces is a homeomorphism (even if P is not projectively faithful, while K is!). This means that $\Lambda_P$ yields a homeomorphism between $d_eS(R_K,1)$ and K. We have proved:

IV.1 **THEOREM** Let K be an integral polytope, and let P be an element of $\mathbf{R}[x_i^{\pm 1}]^+$ such that $\operatorname{cvx} \operatorname{Log} P = K$. Define the function $\Lambda_P: d_eS(R_K,1) \longrightarrow K$ by means of $\Lambda_P(\gamma) = \sum \lambda_w \gamma(x^w/P)\, w$, where $P = \sum \lambda_w x^w$. Then $\Lambda_P$ is a homeomorphism (onto), and it sends the unfaithful pure states of $R_K$ onto the boundary of K.

Finally, we can prove the theorem given in the title of this section. We really only use IV.1 to show that the pure state space of $R_K$ is contractible (being homeomorphic, even diffeomorphic on the interiors, to a convex polytope), and thus has trivial K-theory.

**IV.2 THEOREM** For $S$ any finite subset of $\mathbf{Z}^d$, all finitely generated projective $R_S$-modules are free (with $S = K \cap \mathbf{Z}^d$, we obtain freeness of projective $R_K$-modules).

Proof: Since $d_eS(R_S,1) = d_eS(R_P,1)$, for any $P$ in $(\mathbf{R}^d)^{++}$ with $\operatorname{Log} P = S$, we have by the result (or rather its proof) above, that $X = d_eS(R_S,1)$ is contractible. Thus all finitely generated projective $C(X,\mathbf{R})$-modules are free. Now the natural map $R_S \longrightarrow C(X,\mathbf{R})$, given by $f \mapsto f^\wedge$, where $f^\wedge(\gamma) = \gamma(f)$, embeds $R_S$ as a dense (point-separating!) subring of $C(X,\mathbf{R})$, with the additional property that if $f$ in $R_S$ becomes invertible in $C(X,\mathbf{R})$, then it is up to sign an order unit, so is invertible in $R_S$. This is enough to guarantee that the inclusion of rings induces an embedding on the semigroups of isomorphism classes of finitely generated projective modules [Sw; Theorem 3.1]. As $X$ is contractible, all of its projectives are free and we are done.∎

If we consider, instead of $P$ belonging to $\mathbf{R}[x_i^{\pm 1}]^+$, $P$ belonging to $\mathbf{Z}[x_i^{\pm 1}]^+$, and form the $\mathbf{Q}$-algebra obtained by inverting all the order units in $\mathbf{Z}[x^w/P;\ w \in \operatorname{Log} P]$, $R_{S,\mathbf{Z}}$ ($S = \operatorname{Log} P$), then the theorem above applies to $R_{S,\mathbf{Z}}$ as well. Moreover, $R_{S,\mathbf{Z}}$ itself can be obtained as $K_0(A^{\mathbf{T}})$ for a product type action of the d-torus $\mathbf{T}$ on a UHF C*-algebra $A$. In this case, an immediate consequence of the result above is the rather odd formula

$$K_0((K_0(A^{\mathbf{T}})) = \mathbf{Z}.$$

If however, the localization involved in inverting all the order units is omitted, then freeness of the finitely generated projectives is lost. For example, if $\operatorname{Log} P \subset \mathbf{Z}$ (i.e., $d = 1$) and $P$ is solid, then $R_P$ is integrally closed, finitely generated and has Krull dimension one—so is a Dedekind domain. If $P$ in $\mathbf{R}[x^{\pm 1}]^+$ has no real roots, then the class group has order 2; otherwise $R_P$ is a principal ideal domain. If $P$ is in $\mathbf{Z}[x^{\pm 1}]^+$ and is solid, then $R_P \otimes \mathbf{Q}$ is Dedekind, with class group cyclic of order

$$\gcd\{\text{degrees of factors of } P \text{ in } \mathbf{Z}[x^{\pm 1}]\},$$

and the generator of the class group is the maximal ideal given as the kernel of the pure state $s \mapsto \lim_{t\to\infty} s(t)$. In more variables, just what $K_0(R_P)$ looks like is somewhat mysterious, unless K is "integrally simple"; computations are done in Appendix A, via the Picard group.

## V STATES ON IDEALS

In this section, we provide much of the preparation for subsequent results. We require a number of results about order ideals that are principal, and what their generators can possibly look like.

The rings $R_P$, $R_K$ that we have been discussing admit the following properties that general commutative partially ordered rings may satisfy:

A. 1 is an order unit.

B. The pure state space, $d_eS(R,1)$, is compact [H1; I.2] and connected [HI; III.5].

C. R is noetherian.

D. R admits interpolation and is unperforated (for $R_P$ and $R_K$, these properties hold by construction).

E. R is a domain ($R_P$ and $R_K$ are subrings of $\mathbf{R}(x_i)$).

F. For x in $R^+\setminus\{0\}$, $\{L \in d_eS(R,1) \mid L(x) = 0\}$ is nowhere dense in $d_eS(R,1)$.

To see that F holds for $R_P$ and $R_K$, regard either ring as a ring of rational functions. The evaluations at points $r = (r_1,...,r_d) \in (\mathbf{R}^d)^{++}$ define pure states of the ring, and by [H1; III.5], these are dense in the pure state space. However, point evaluations cannot kill a nonzero positive element.

If I is an order ideal of a partially ordered ring R satisfying A through F above, then I admits an order unit (see the comments in [H2]). Even though the pure state space of R is connected, it is not generally true that the pure state space of I is connected. Thus the following result is not completely obvious:

**V.1 THEOREM** Assume R is a commutative partially ordered ring satisfying A through F above. Let I be an order ideal of R, and suppose that $I = zR$ for some z in R. Then either z or $-z$ is an order unit of R (so in particular, either z or $-z$ lies in $R^+$), and if for some element u in R, $y = zu$ is an order unit for I, then u is an order unit of R.

Note that property A implies that an order ideal is an ideal [H1; I.2(a)], and C implies that all order ideals have order units (viewing the order ideal as a partially ordered abelian group. In order to prove the theorem, we require a number of lemmas, some of which are interesting in their own right.

**V.2 LEMMA** (c.f., [GH2; 3.1]) Let $(G, u)$ be a partially ordered abelian group with order unit, $u$. If $f$ is a state of $(G, u)$ such that

for all *order units* $x,y$ of $G$,

$$\sup\{f(z) \mid 0 \leq z \leq x,y\} = \min\{f(x),f(y)\},$$

then $f$ is a pure state.

Proof: The proof parallels that of [GH2; 3.1]. Suppose $f = \alpha g + (1-\alpha)h$, where $g$ and $h$ are distinct states of $(G, u)$, and $\alpha$ is a real number in the open interval $(0,1)$. Then $g|G^+ \neq h|G^+$ (as $G=G^+-G^+$), and as $G^{++} \supset Nu + G^+$, (the latter denotes the set of order units), it follows that $g|G^{++} \neq h|G^{++}$. There thus exists $a$, an order unit of $G$, such that $g(a) < h(a)$. There exist positive integers $m$ and $n$ so that

$$g(a) < m/n < h(a).$$

Set $x = na$, $y = mu$. Then $g(x) < g(y) = m$, and $h(x) > h(y)$. If $0 \leq z \leq x, y$, then

$$0 \leq f(z) = \alpha g(z) + (1-\alpha)h(z) = f(x)-(1-\alpha)(h(x)-h(y))$$

$$\leq \alpha g(x) + (1-\alpha)(h(y))$$

$$\leq f(y) - \alpha(g(y)- g(x)).$$

Thus $\sup\{f(z) \mid 0 \leq z \leq x, y\}$ is strictly less than $\min\{f(x),f(y)\}$, a contradiction.∎

The following is a partial converse to [H1;I.2(c)]:

**V.3 LEMMA** Suppose the partially ordered ring $R$ satisfies properties A, D, E, F. Then, given $L_o$ in $d_eS(R,1)$, there exists $B$ in $d_eS(I,u)$ ($u$ a fixed but arbitrary order unit for the order ideal $I$), so that for all $s$ in $R$, and $i$ in $I$,

$$\beta(si) = L_o(s)\beta(i).$$

Proof: Set $K = Ru$, with the ordering given by $K^+ = K \cap R^+$; obviously $I \supset K$, and although $K$ is an ideal, it is not an order ideal (unless $K = I$). There is a well-defined map from $R$ to $K$, given by $r \mapsto ru$.

This is order-preserving (but not necessarily an order-isomorphism). Define $\alpha:(K,u) \longrightarrow (R,1)$ via $\alpha(ru) = L_o(r)$. Then $\alpha$ is well-defined, as $R$ is a domain. If $ru \geq 0$, then $L_o(ru) \geq 0$; as $L_o(ru) = L_o(r)L_o(u)$, and $L_o(u) \geq 0$, we deduce that $L_o(r) < 0$ would imply $L_o(u) = 0$. By F applied to $ru = x$, $L(r) > 0$ for a dense set of $L$'s in $d_eS(R,1)$, so $L(r) \geq 0$ for all $L$, and thus $L_o(r) \geq 0$. As $\alpha$ is obviously additive, $\alpha$ is a state of $(K,u)$.

Now we show that $\alpha$ is a pure state of (K,u). We use the criterion of the previous lemma. Pick order units of (K,u), x = su, y = tu, with s and t in R. There exists a rational number d exceeding zero so that su and tu are greater than or equal to du (if R is not at least a $\mathbf{Q}$-algebra, it may be replaced by $R \otimes \mathbf{Q}$). Hence $(s-d)u,(t-d)u \geq 0$. Thus for all pure states L of (R,1), $L(s-d), L(t-d) \geq 0$ (by the result of the preceding paragraph).

Hence $L(s), L(t) \geq d > 0$ for all L in $d_eS(R,1)$, so s and t are order units in R (by A and D). By the full extremal criterion of [GH2; 3.1(c)], given $\varepsilon > 0$, there exists v in R such that

$$0 \leq v \leq s,t \quad \text{and} \quad L(v) > \min\{L(s), L(t)\} - \varepsilon.$$

Thus $\alpha(vu) > \min\{\alpha(su), \alpha(tu)\} - \varepsilon$, and as $0 \leq vu \leq su, tu$, the purity criterion of the previous lemma is satisfied. Thus $\alpha: K \longrightarrow \mathbf{R}$ is a pure state.

Since u is also an order unit for I, $\alpha$ extends to a state on I [GH1;4.1] (as $K^+ = K \cap I^+$), and since $\alpha$ is pure, we may extend it to a pure state $\beta$ of (I,u).

By property A and [H1;I.2], there exists a pure state L' of R such that for all s in R and i in I, $\beta(si) = L'(s)\beta(i)$. Setting i = u, we obtain that $L'(s) = \beta(su) = \alpha(su) = L(s)$, so $L' = L_o$, and we are done.∎

Proof of V.1: By C, I admits an order unit, call it y. There exists r in R such that y = rz. Let $\beta$ be a pure state of (I,y). By [H1;I.2(c)], there exists $L_1$ in $d_eS(R,1)$ such that either $L_1(y)\beta = L_1|I$ ($L_1$ restricts to $\beta$), or $L_1(y) = 0$ and $1 = \beta(y) = L_1(r)\beta(z)$. In either case, $L_1(r)\beta(z) = 1 \neq 0$, so $L_1(r) \neq 0$. Replacing r, z by −r, −z, we may assume if necessary that for some L in $d_eS(R,1)$, that $L(r) > 0$. By V.3, $L_o(r) \neq 0$ for all $L_o$ in $d_eS(R,1)$, so that by the connectedness assumption (B), we deduce that for all $L_o$ in $d_eS(R,1)$, $L_o(r) > 0$. Hence r is an order unit for R. Also, $\beta(z) = 1/L_1(r)$, so $\beta(z) > 0$ for all $\beta$ in $d_eS(I,y)$. Thus z is an order unit for I.∎

**REMARK:** Connectedness of $d_eS(R,1)$ does not imply connectedness of $d_eS(I,u)$ (even when the latter is compact). For example, set $R = \mathbf{Z}[X,Y]$, with positive cone $R^+$ generated multiplicatively and additively by {X, Y, 1−X−Y} (this is $R_P$ with P = 1+x+y, d = 2). Let M = (X,Y); this is a prime ideal and an order ideal. Set $I = M^2 = (X^2, XY, Y^2)$; this is an order ideal. All pure states of I with a positive element in the kernel kill $M^3$, (this follows easily from [H1; I.2(c)] and a little manipulation), and $I/M^3 \approx \mathbf{Z}^3 \approx \underline{X}^2\mathbf{Z} \oplus \underline{XY}\mathbf{Z} \oplus \underline{Y}^2\mathbf{Z}$ (where the underlined terms represent the images modulo $M^3$) with the coordinatewise ordering. The projection onto the $\underline{XYZ}$ component is a pure state, and it is easy to see that this pure state is an isolated point in $d_eS(I,u)$, with $u = X^2 + XY + Y^2$.∎

The following is useful, as it shows that prime ideals that are minimal with respect to lying over order ideals are themselves order ideals:

**V.4 PROPOSITION** Let $I$ be an order ideal in some $R_P$. If $\mathfrak{p}$ is a prime ideal of $R_P$ containing $I$, then there exists an order ideal $\mathfrak{q}$ of $R_P$ that is prime and satisfies $\mathfrak{p} \supset \mathfrak{q} \supset I$.

Proof: If $I = (0)$, then $\mathfrak{q} = (0)$ will do. Otherwise let $\mathfrak{q}$ be the ideal generated by $\{x^w/P \mid x^w/P \in \mathfrak{p}\}$, and set $Z = \{w \in \text{Log } P \mid x^w/P \in \mathfrak{p}\}$. Notice that $Z$ is not empty, since $I$ is generated by products of terms of the form $x^w/P$, and $\prod r_i \in \mathfrak{p}$ implies at least one of the $r_i$ belongs to $\mathfrak{p}$. By II.2A, $\mathfrak{q}$ is an order ideal. We claim $Z = \text{Log } P \backslash F$ for some face $F$ of $\text{cvx Log} P$.

To this end, observe (as in the proofs of III.1 and III.2, for example) that $\{x^v/P \mid v \in d_e \text{Log} P\}$ generates the improper ideal. There thus exists $v$ in $d_e\text{Log } P$ such that $x^v/P$ does not belong to $\mathfrak{p}$, i.e., $v \notin Z$.

Set $Y = \text{Log } P \backslash Z$, and let $S = \text{cvx } Y$. We wish to show $S$ is a face of cvx Log P.

***(a) If $w \in S$ and $Nw \in N\text{Log } P$ (sums of $N$ elements from Log P) for some integer $N$, then $x^{Nw}/P^N$ does not belong to $\mathfrak{p}$.***

Write $w = \sum \alpha_i v_i$, $\alpha_i > 0$, $v_i \in Y$, $\sum \alpha_i = 1$. By [HI;III.1A], we may assume that there exists a positive integer $M$ such that all of the $M\alpha_i$ are integers. Hence $NMw = \sum (MN\alpha_i) v_i$, and we have

$$\left(\frac{x^{Nw}}{P^N}\right)^M = \frac{x^{MNw}}{P^{MN}} = \prod_i \left(\frac{x^{v_i}}{P}\right)^{NM\alpha_i}.$$

As none of the $x^{v_i}/P$ belong to $\mathfrak{p}$, and the latter is prime, $(x^{Nw}/P^N)^M$ does not belong, so neither does $x^{Nw}/P^N$.

***(b) (i) If $w \in Q^d$ and $Nw \in N\text{Log } P$, $\underline{\text{and}}$ $w \notin S$, then $x^{Nw}/P^N \in \mathfrak{p}$.***

***(ii) If $Nw \in N\text{Log } P$, $w = \alpha v + (1-\alpha)v'$ (where $0 < \alpha < 1$), $v' = \text{cvx Log } P$, and $v \in (\text{cvx Log} P)\backslash S$, then $x^{Nw}/P^N \in \mathfrak{p}$, and $w \notin S$.***

(To (i)). Write $w = \sum \alpha_j e_j$ where $\alpha_j > 0$ and $\sum \alpha_j = 1$, and in addition, $e_j \in d_e\text{Log } P$; as above, we may assume the $\alpha_j$ are all rational. If all the $e_j$'s belong to S, then w belongs to S. There thus exists f in $\{e_j\} \backslash S$. Write $w = \alpha f + (1-\alpha)g$, with $g \in (\text{cvx Log } P) \cap Q^d$, and $0 < \alpha < 1$ with $\alpha$ in **Q**. There exists a positive integer N so that $M\alpha$ is an integer and so that Mf and Mg belong to MLog P (this is possible, since the original $e_j$'s lay in Log P). From

$$NM^2 w = N(M\alpha)Nf + NM(1-\alpha)Mg,$$

we obtain

$$\left(\frac{x^{Nw}}{P^N}\right)^{M^2} = \left(\frac{x^{Mf}}{P^M}\right)^{NM\alpha} \cdot \left(\frac{x^{Mg}}{P^M}\right)^{NM(1-\alpha)} \in \mathfrak{p}.$$

The last portion follows from $x^f/P$ belonging to $\mathfrak{p}$. As $\mathfrak{p}$ is prime, $x^{Nw}/P$ lies in $\mathfrak{p}$, as desired.

(To (ii)). If $w \notin S$, then (i) applies. So we assume w belongs to S and deduce a contradiction. Write $v = \sum \alpha_i f_i$, with $f_i$ in $d_e\text{Log } P$, not all of the $f_i$'s in S; $\alpha_i$ in $Q^+ \backslash \{0\}$, and $\sum \alpha_i = 1$. Also write $v' = \sum a_j g_j$, with $g_j$ in $d_e\text{Log } P$, $a_j$ in $Q^+ \backslash \{0\}$, and $\sum a_j = 1$. We can then express w as

$$w = \sum \beta_k f_k' + \sum b_m g_m'$$

where the $f_k'$ belong to $d_e\text{Log } P \setminus S$, the $g_m'$ belong to $S \cap \text{Log } P$, and $\beta_k$, $b_m$ are strictly positive rational numbers with $1 = \sum \beta_k = \sum b_m$. There exists an integer M so that all of $M\beta_k$ and $Mb_k$ are integers. Hence

$$\left(\frac{x^{Nw}}{P^N}\right)^{M} = \prod \left(\frac{x^{f_k'}}{P}\right)^{NM\beta_k} \cdot \prod \left(\frac{x^{g_m'}}{P}\right)^{NMb_m}$$

and this last expression belongs to $\mathfrak{p}$, as each $x^{f_k'}/P$ does. Thus $x^{Nw}/P^N$ belongs to $\mathfrak{p}$, which contradicts (a) and w belonging to S. ,

*(c) $S \cap \text{Int } K$ is empty, where $K = \text{cvx Log } P$.*

Assume the intersection is not empty. Since S is the convex hull of lattice points, its rational points are dense in it—so every ball (of full dimension) inside S contains a rational point. Hence there exists s in $Q^d \cap \text{Int } K \cap S$. There exists w in $d_e\text{Log } P \backslash S$, and the line joining w to s can be extended slightly (as s lies in Int K), that is, we may write $s = \alpha w + (1-\alpha)w'$ where w' belongs to K and $1 > \alpha > 0$. This line segment may be extended to a rational point (as both s and w are rational), and so we may assume that $w$ is a rational point in $\mathrm{Int}\, K \cap S$. Then $\alpha$ is rational, so by (b)(ii) $s$ does not belong to $S$, a contradiction.

***(d) There exists a proper face T of K containing S.***

We have that $S$ is a convex polyhedron inside $K$, and $\mathrm{Int}\, K \cap S$ is empty. Thus $S$ is contained in the boundary of $K$ which is a union of faces. As $S$ is convex, $S$ is contained in a single face of $K$.

***(e) Conclusion of the proof:***

By (d), $S$ is contained in one of the proper faces $T$ of $K = \mathrm{cvx}\,\mathrm{Log}\, P$. There exists a "relative" face $H$ of $\mathrm{Log}\, P$ corresponding to $T$, that is, $\mathrm{Log}\, P \cap T = H$ and $\mathrm{cvx}\, H = T$. Form the quotient ring $R_{P_H} = R_P/\langle\{x^w/P \mid w \in \mathrm{Log}\, P \setminus H\}\rangle$ where $P = \sum \lambda_w x^w$, and $P_H = \sum_{w \in H} \lambda_w x^w$. As $H$ is proper, $\dim T = \dim K$. By induction on dimension, $S$ is a face of $T$, hence of $K$. Thus $S = \mathrm{cvx}\, Y$ is a face of $K$. We claim $S \cap \mathrm{Log}\, P = Y$. If $w$ belongs to $S \cap \mathrm{Log}\, P$, then (a) applies with $N = 1$, so that $v \in Y$. By [H1; §VII], $\mathfrak{q}$ is prime. ∎

## VI. FACTORIALITY AND INTEGRAL SIMPLICITY

Let v be a vertex of an integral polytope K. Let $L_1, \ldots, L_d$ be a list of the 1-dimensional faces (edges) of K that contain v. Let $v_i$ be the nearest lattice point to v along $L_i$ for each i = 1,..., t, and set $K_v = \mathrm{cvx}\{v, v_1, \ldots, v_t\}$. Obviously $K_v$ is a subpolytope of K and is of the same dimension. We refer to $K_v$ as the peak polytope of K at v. Many ring–theoretic properties of $R_P$ and $R_K$ relate to the geometric behaviour of the peak polytopes. In this section, we shall show that if K is projectively faithful, then $R_K$ is factorial if and only if each peak polytope (one for each vertex of K) is AGL(d,Z)-equivalent to the standard solid simplex of the same dimension as K (equivalently, when K is d-dimensional, each peak polytope has d-volume exactly 1/d!). This is also characterized more geometrically.

The idea of the proof is that for suitable P, $R_P$ is (homologically) regular, and the freeness of all projective $R_K$-modules (§IV) yields unique factorization.

We require a special case of a definition from [H2;section I]. Let v be a vertex of K, and E a facet of K containing v. Set $S_v = (K \setminus E) \cap \mathbf{Z}^d$. A singleton subset {s} of $S_v$, is called a dominant stratum of K with respect to v, if whenever $s \in n(K \cap \mathbf{Z}^d) + z$ for some positive integer n, and z in $\mathbf{Z}^d$ with $s \neq nv+z$, then $nv+z \notin S$. With K = cvx Log P and {v} = F, this corresponds to the (more general) definition given in [H2]. We say that K has the unique dominant stratum property if for each facet E (of K) and vertex v thereof, $S_v$ has a unique dominant stratum with respect to v.

Finally, we say the integral d–dimensional polytope K in $\mathbf{R}^d$ is weakly integrally simple if it satisfies the following two properties:

(a) Viewed as a real polytope, K is simple; i.e., each peak polytope is a simplex.

(b) For each vertex v of K, and facet E containing v, let L be the unique (by (a)) edge of K such that $L \cap E = \{v\}$, and let $\alpha$ be the linear functional defining E, say $E = \alpha^{-1}(r) \cap K$ for some real r, and $\alpha|K \geq r$. Let w be the nearest lattice point on L to v. Then there are no lattice points between E and E+w–v, in the sense that if z belongs to $nK \cap \mathbf{Z}^d$, then $\alpha(w) > \alpha(z) > \alpha(v)$ is impossible.

We recall from section III the definition of integrally simple—at each vertex, the cone generated is simplicial, or in other words the the fundamental polytope of the cone is AGL(d,Z)-equivalent to the standard simplex. In other words, K is integrally simple if and only if each peak polytope is AGL(d,Z)–equivalent to the standard simplex.

We would like to prove the following:

**THEOREM(?)** Let K be a projectively faithful integral polytope in $\mathbf{R}^d$. The following are equivalent:

(i) $R_K$ is factorial.

(ii) K has the unique dominant stratum property.

(iii) K is weakly integrally simple.

(iv) K is integrally simple.

We would *like* to, but unfortunately difficulties arise from the open question (section III) as to whether K projectively faithful and $R_K$ integrally closed imply that K is solid. We will however show that (i) implies (ii), which in turn implies (iii), and (iv) implies (i) is easy. Then properties (ii) and (iii) will be strengthened as follows:

(ii') Both K and dK have the unique dominant stratum property.

(iii') Both K and dK are weakly integrally simple.

The characterization thereupon becomes complete: That (i) implies (ii') follows from (i) implies (ii) and the fact that $R_K = R_{dK}$ when $R_K$ is integrally closed and K is projectively faithful; (ii') implies (iii') follows from (ii) implies (iii), and now (iii') implies (iv) is relatively straightforward. Note that if $d = 2$, then (ii'), (iii') are equivalent to their respective counterparts, so the theorem as originally stated can be proved. The main result of this section will be:

VI.1 **THEOREM** For K a projectively faithful integral polytope in $\mathbf{R}^d$, the following are equivalent:

(i) $R_K$ is factorial.

(ii') Both K and dK have the unique dominant stratum property.

(iii') Both K and dK are weakly integrally simple.

(iv) Every peak polytope of K is of volume $1/d!$

Furthermore, if K is locally solid, then (ii') and (iii') may be replaced by (ii) and (iii) respectively.

Of course, (iv) is just the definition of integral simplicity. At this point, let us examine some of the examples. If $P = \sum \lambda_i x_i + \lambda_o$, or if $P = \prod (1+\lambda_i x_i)$, then K is respectively the standard d-simplex, the standard d-cube, so in both cases K is integrally simple. Obviously $R_K$ is a localization of a pure polynomial ring, and so has unique factorization.

If $P = 1 + x + y + y^2$, then $K = \mathrm{cvx}\{(0,0),(1,0),(0,2)\}$ and this is projectively faithful; but not integrally simple. If the vertex v is either (0,0) or (0,2), then $K_v$ is the standard unit triangle, but if $v = (0,1)$, $K_v$ is just K itself, and this is not integrally simple; in fact by VIII.15, any integrally simple simplex is an integer multiple of a ***standard*** simplex.

To compute the relevant dominant strata, set $E = \{(0,2),(0,1)\}$ and $v = \{(0,2)\}$. Then E is the intersection of a facet of K with $\mathbf{Z}^d$. Set $s_1 = (1,0)$, and $s_2 = (0,0)$; we shall show that each of $\{s_1\}$, $\{s_2\}$ is a dominant stratum with respect to v. Suppose that $s_1 = \sum_{1 \le i \le n} k_i + z$ for $k_i$ in $K \cap \mathbf{Z}^d$ and z in $\mathbf{Z}^d$. If $n(0,2) + z \in S_v = \{s_1,s_2\}$ and we assume that $s_1 \neq nv + z$, then $s_2 = n(0,2) + z$, and so $z = (0,-2n)$. Thus $(1,0) = \sum k_i - (0,2n)$. As at least one of the $k_i$'s must be (1,0) and all the rest are forced to be of the form $k_i = (0,t_i)$ for $t_i$ in $\{0,1,2\}$, and $0 = \sum t_i - (0,2n)$, we obtain a contradiction. Thus $\{s_1\}$ is a dominant stratum.

Now suppose $s_2 = \sum_{1 \le i \le n} k_i + z$ and $s_1 = (0,2n) + z$. We deduce that $z = (-1,-2n)$, so that $(0,0) = \sum k_i - (1,2n)$. Hence as in the previous paragraph, at least one of the $k_i$'s must be (1,0) and all the rest are forced to be of the form $k_i = (0,t_i)$ for $t_i$ in $\{0,1,2\}$, and again a contradiction results from $\sum t_i < 2n$.

Thus we have shown directly that this particular choice for K fails to have the unique dominant stratum property.∎

VI.2 **LEMMA** (Ad (i) implies (ii)). Let K be an integral polytope in $\mathbf{R}^d$, define $P = \sum_{K \cap \mathbf{Z}^d} \lambda_w x^w$ of $\mathbf{R}[x_i^{\pm 1}]^+$, and let F be a facet of K (= cvx Log P). Let v be a vertex of F, and set $\mathfrak{q} = \mathfrak{q}_F$ to be the ideal of $R_P$ generated by $\{x^w/P \mid w \in (K \cap \mathbf{Z}^d) \setminus F\}$. Then $\mathfrak{q}$ ($\mathfrak{q}R_K$) is a minimal prime ideal of $R_P$ ($R_K$) respectively, it is an order ideal, and if there are at least two dominant strata of $S = (K \cap \mathbf{Z}^d) \setminus F$ with respect to $\{v\}$ as a relative face of $K \cap \mathbf{Z}^d$, then $\mathfrak{q}$ ($\mathfrak{q}R_K$) is not even projective, let alone principal, as an ideal.

Proof: Call two of the dominant strata (with respect to the same vertex v of K), v', v". By [H1;III.3], there exists a pure state $L: R_P \to \mathbf{R}$ corresponding to v, that is, $L(x^v/P) = 1$, $L(x^w/P) = 0$ for $w \in K \cap \mathbf{Z}^d \setminus \{v\}$. By [H2; §IV], there exist pure states β' and β" of $\mathfrak{q}$ such that $\beta'(x^{v'}/P) \neq 1$, $\beta'(x^w/P) = 0$ for $w \in K \cap \mathbf{Z}^d \setminus \{v,v'\}$, and similarly with β". By [H2;I.2(c)], (or by the construction given in [H2;§IV]), for r in $R_P$, and s in $\mathfrak{q}$, $\beta'(rs) = L(r)\beta'(s)$, and similarly with β". If $\mathfrak{q}$ were projective, then at every localization at a maximal ideal containing it, say M, $\mathfrak{q}(R_P)_M$ would be principal. We shall select a maximal ideal M which is also an order ideal; then M will contain no order units, so that $(R_P)_M$ will also be a localization of $R_K$. We will show that $\mathfrak{q}(R_P)_M$ cannot be principal.

Choose for our M the kernel of L; this is an order ideal by [H1; §VII], and it is certainly maximal as an ideal. Moreover, it is generated as an ideal by $\{x^w/P \mid w \in (K \cap \mathbf{Z}^d) \setminus \{v\}\}$. Suppose $\mathfrak{q}(R_P)_M$ were principal. Then there would exist t in $R_P$ such that $(x^{v'}/P)\,a' = t\,s'$ can be solved with s' in $R_P$ and a' in $R_P$ with $L(a') \neq 0$, and similarly there would exist s" and a" in $R_P$ with $L(a'') \neq 0$ such that $(x^{v}/P)\,a'' = t\,s''$.

Apply $\beta'$ to the first equation; we obtain

$$\beta'(x^{v'}/P)\,L(a') \;=\; \beta'(t)\,L(s').$$

Since both terms on the left side are not zero, we deduce that $\beta'(t)$ is not zero, and also $L(s')$ is not zero. Apply $\beta''$ to the first equation; we obtain (as $\beta''(x^{v'}/P) = 0$), that $\beta'(t)\,L(s')$ is zero, and so $\beta''(t)$ is zero. However, if we apply $\beta''$ to the second equation, $((x^{v''}/P)\,a'' = t\,s'')$, we deduce as in the first case that $\beta''(t)$ is not zero, a contradiction.∎

To prove (ii) implies (iii) requires a few lemmas:

**VI.3 LEMMA** Let K be an integral polytope, v a vertex of K, and F a facet of K; let L be an edge (1–face) of K such that $L \cap F = \{v\}$. If v' is the nearest lattice point to v on L, then $\{v'\}$ is a dominant stratum of K with respect to v.

<u>Proof</u>: Suppose that v' lies in $n(K \cap \mathbf{Z}^d) + z$ for some positive integer n and lattice point z, $v' \neq nv+z$, and there exists s in $K \cap \mathbf{Z}^d \setminus F$ with $s = nv+z$. Write $v' = \sum_{1 \le j \le n} v_j + z$ with $v_j$ in $K \cap \mathbf{Z}^d$; we obtain

$$s + \textstyle\sum_{1 \le j \le n} v_j \;=\; v' + nv \qquad\qquad (*).$$

Since v and v' lie on the 1–face L, there exist linear functionals $\beta_i$ and corresponding real numbers $k_i$ with

$$K \cap \left( \cap \beta_i^{-1}(k_i) \right) = L \qquad \text{where} \quad \beta_i|K \ge k_i.$$

Applying the $\beta_i$'s to (*), we obtain

$$\beta_i(s) + \sum_j \beta_i(v_j) \;=\; \beta_i(v') + n\,\beta_i(v) \;=\; (n+1)\,\beta_i(v').$$

As each of $\beta_i(s)$, $\beta_i(v_j)$ is less than or equal to $k_i = \beta_i(v')$, we obtain $\beta_i(s) = \beta_i(v')$ for all i. Thus s belongs to L. We may therefore write $s = v + t(v'-v)$ with t a real number exceeding one (as v' is the nearest lattice point to v along L). If $F = \alpha^{-1}(r) \cap K$ and $\alpha|K \geq r$ for a linear functional $\alpha$, applying it to (*) yields

$$\alpha(s) + \sum_j \alpha(v_j) = \alpha(v') + n\,\alpha(v);$$

from $\alpha(s) \geq \alpha(v')$ and $\alpha(v_j) \geq \alpha(v)$, we deduce that these inequalities are actually equalities, so $\alpha(s) \geq \alpha(v')$. Hence $\alpha(s) = \alpha(v) + t\alpha(v') - t\alpha(v)$; we have $(1-t)\alpha(v') = (1-t)\alpha(v)$, so $\alpha(v) = \alpha(v')$, a contradiction.∎

VI.4 **LEMMA** If the integral polytope K has the unique dominant stratum property, then it is simple as a real polytope.

Proof: If K were not simple, there would be a vertex v with more than d 1-faces emanating from it. There exists a facet F of K containing v that has two 1-faces, $L_1$ and $L_2$ with $L_i \cap F = \{v\}$. Each nearest point on $L_i$ is a dominant stratum of K with respect to v, contradicting the hypotheses.∎

Conclusion of the proof that (ii) implies (iii): Part (a) of the definition of weakly integrally simple having been established above, we now work toward proving that part (b) holds. To this end, let F be a facet of K, $\alpha$ a functional with $\alpha|F = k$, $\alpha|K \leq k$, v a vertex of F, and v' the nearest lattice point to v along a 1–face L not contained in F. Write $\alpha(v') = k+m$ with $m > 0$.

If there exists s in $K \cap \mathbf{Z}^d$ such that $k < \alpha(s) < k+m$, select $s_o$ in $K \cap \mathbf{Z}^d$ such that $\alpha(s_o)$ is minimal among those values greater than k, say $\alpha(s_o) = k+m_o$. Let $\beta_1$ be one of the linear functionals describing L as In the proof of VI.2. Select $s_1$ in $K \cap \mathbf{Z}^d$ such that $\beta_1(s_1)$ is minimal among those s's in $\alpha^{-1}(k+m_o) \cap K \cap \mathbf{Z}^d$. Among the lattice points in $\alpha^{-1}(k+s_o) \cap \beta_1^{-1}(\beta_1(s_o)$, select $s_2$ having the smallest $\beta_2$ value; continue the process with the rest of the $\beta_i$'s that describe L; eventually we get down to a singleton {s'} (the intersection of the hyperplanes described by the $\beta_i$'s is L), and this is intersected with the hyperplane $\alpha^{-1}(\{k+m_o\})$.

We claim {s'} is a dominant stratum of K with respect to {v}. Say $s' = \sum_{1 \leq j \leq n} v_j + z$ with $v_j$ in $K \cap \mathbf{Z}^d$ not all equal to v and $f = nv+z$ where f belongs to $K \cap \mathbf{Z}^d \backslash F$. We obtain

$$s' + nv = f + \sum v_j \qquad (*).$$

As in the proof of the previous lemma, we apply $\alpha$ and the $\beta_i$'s. From $\alpha(s') \leq \alpha(f)$, we obtain equality, $\alpha(s') = \alpha(f)$. Apply $\beta_1$; from $\beta_1(s') \leq \beta_1(f)$, (arising from $\alpha(s') = \alpha(f)$), we deduce $\beta_1(s') = \beta_1(f)$. We may continue this process, applying $\beta_2, \beta_3, \ldots$ and we deduce $s' = f$ (from the uniqueness of $s'$ satisfying the inequalities). Hence $\{s'\}$ is a dominant stratum; as $s' \neq v'$, we have a contradiction to the uniqueness of dominant strata. This yields part (b) of the definition of weakly integrally simple.∎

**VI.5 LEMMA** (Ad (iii') implies (iv)). If $K$ and $dK$ are weakly integrally simple integral polytopes, then each peak polytope of $K$ is AGL(d,Z)–equivalent to the standard solid unit simplex.

<u>Proof</u>: Let $v$ be a vertex of $K$, and let $K_v$ be the corresponding peak polytope. We show first that $K_v$ is a *building block* (that is, a simplex satisfying $K_v \cap \mathbf{Z}^d = d_e K_v$).

By hypothesis, $K_v$ is a simplex, and its vertices other than $v$ will be nearest points to $v$ along edges, say $\{v_1, v_2, \ldots, v_d\}$ with corresponding edges $\{L_1, L_2, \ldots, L_d\}$. Suppose $w$ belongs to $K \cap \mathbf{Z}^d$ with $w$ unequal to any of the elements of $\{v, v_1, v_2, \ldots, v_d\}$. We may write $z = t v + \sum t_i v_i$, where the $t$'s are non–negative real numbers that add to 1. Select a $j$ such that $t_j \neq 0$ (necessarily, $t_j \neq 1$ as well). The set of points $\{v, v_1, v_2, \ldots, \hat{v}_j, \ldots, v_d\}$ is a subset of a unique facet $F$ of $K$, with the affine span of the set equalling that of $F$. Let $\alpha$ be a linear functional such that $K \cap \alpha^{-1}(r) = F$ and $\alpha | K \geq r$.

Now we have that $v_i$ is the nearest point to $v$ along the edge $L_i$, and $L \cap F = \{v\}$ (the former being at most 0–dimensional and the latter belonging to it). We observe that

$$\alpha(z) \quad = \quad r\left(t + \sum_{i \neq j} t_j\right) \quad + \quad t_j \alpha(v_j);$$

the right side is strictly less than $\alpha(v_j)$, contradicting part (b) of the definition of weakly integrally simple. We deduce that no such $z$ exists, so $K_v \cap \mathbf{Z}^d = d_e K_v$. This portion of the proof only required that $K$ be weakly integrally simple.

Next we show that $K_v$ is affinely integrally equivalent to the unit simplex. Notice that $(dK)_{dv}$ (the corresponding peak polytope of $dK$) is just a translation of $K_v$. By subtracting $v$ from everything, we may assume $v = \mathbf{0}$. As $K_v \cap \mathbf{Z}^d = d_e K_v$, after applying a transformation from GL(d,Z), we may assume $v_1 = (1,0,\ldots,0)$. By further such transformations (using the linear independence of the set $\{v_1, v_2, \ldots, v_d\}$) we find that either we may assume $v_j = (0,\ldots,0,1,0,\ldots,0)$ for $1 \leq j \leq d$, in which case we are done, or there exists $k \leq d-1$ such that $v_i = (0,\ldots,0,1,0,\ldots,0)$ (1 in the i-th position) for all $i \leq k$ and $v_{k+1} = (a_1, \ldots, a_k, n, 0, \ldots, 0)$ for some positive integer $n$ and integers $a_1, \ldots, a_k$. Subtraction of the $k+1$–st entry from any other is an elementary operation (implemented by an element of GL(d,Z), so we may assume that $0 \leq a_i < n$. Let $F$ be the (unique) facet containing $\{v, v_1, v_2, \ldots, \hat{v}_{k+1}, v_{k+1}, \ldots, v_d\}$; the corresponding linear functional $\alpha$ satisfies $\alpha|F=0$, and $\alpha|K \leq 0$. Moreover $v_{k+1}$ is the nearest point to $v$ along some edge $L$ with $L \cap F = \{v\}$. Define

$$w = \frac{1}{n} v_{k+1} + \sum_{i \leq k} \left(1 - \frac{a_i}{n}\right) v_i;$$

we see immediately that

$$w = (1,1,\ldots,1,0,0,\ldots,0) \in \mathbf{Z}^d.$$

The sum of the coefficients in this expansion of $w$ is $1/n + k - (1/n)\cdot\sum a_i$; this is strictly less than $k$ (if all of the $a_i$ are zero and $n > 1$, then $K_v$ would not be a building block), so $w$ belongs to $kK$, hence to $dK$. Moreover, $\alpha(w) = (1/n)\alpha(v_{k+1})$ (as $\alpha(v_i) = 0$ for $1 \leq i \leq k$). Since we have assumed that $v$ is zero, $v_{k+1}$ is a nearest lattice to $v = dv$ along an edge of $K$, and $\alpha(v_{k+1}) > \alpha(w) > \alpha(dv) = 0$, contradicting $dK$ being weakly integrally simple.∎

**REMARK:** The following allows slightly more general forms for the choice of $P$, in that $\operatorname{Log} P = K \cap \mathbf{Z}^d$ is not strictly necessary. It turns out that the condition cited is necessary as well. Moreover, the analogue of the III.2 is also true: If $K$ is a locally solid projectively faithful integral polytope and $P$ in $\mathbf{R}[x_i^{\pm 1}]^+$ satisfies

$$\bigcup_{v \in d_e K} (K_v \cap \mathbf{Z}^d) \subset \operatorname{Log} P \subset K$$

then $R_P$ is integrally closed (and an order in $R_K$). More generally, it should be true that if $K = \operatorname{cvx} \operatorname{Log} P$, is projectively faithful, then $R_P$ is an order in $R_K$ if and only if the same condition holds.

VI.6 **PROPOSITION** Let $K$ be a projectively faithful integral polytope which is integrally simple. Let $P$ in $\mathbf{R}[x_i^{\pm 1}]^+$ be such that

$$\bigcup_{v \in d_e K} (K_v \cap \mathbf{Z}^d) \subset \operatorname{Log} P \subset K$$

(that is, $\operatorname{cvx} \operatorname{Log} P = K$ and $\operatorname{Log} P$ contains all of the lattice points in all of the peak polytopes of $K$). Then every localization of $R_P$ (hence of $R_K$) at a prime ideal $\mathbf{p}$ is isomorphic to a localization of the pure polynomial algebra $\mathbf{R}[x_i]$, and so each localization is regular and factorial. Additionally, $R_P$ is integrally closed, regular, and all minimal prime ideals of $R_P$ are projective.

Proof: Since $\{x^v/P | v \in d_eK\}$ generates an improper ideal of $R_P$ (as we have seen many times previously), there exists $v$ in $d_eK$ such that $x^v/P \notin \mathfrak{p}$. Hence $(R_P)_{\mathfrak{p}}$ is a localization of $R_P[(x^v/P)^{-1}]$; assuming $v = \mathbf{0}$ and $K \subset (\mathbf{R}^d)^+$ (as we may by a translation and an application of an element of $GL(d,\mathbf{Z})$), $R_P[(x^v/P)^{-1}]$ becomes $\mathbf{R}[x_i, P^{-1}]$, a localization of the pure polynomial ring in $d$ variables. This is of course regular and factorial. The other statements follow as we are localizing at an arbitrary prime ideal.∎

VI.7 **LEMMA** (Ad(iv) implies (i)). Let K be an integrally simple, integral polytope. Then $R_K$ is factorial and regular.

Proof: It is clear that integral simplicity implies projective faithfulness, so VI.6 applies to $R_K$. By IV.2, all projectives over $R_K$ are free, so all minimal prime ideals of $R_K$ are principal. As $R_K$ is noetherian, [K; Theorem 178, p. 132] applies, yielding that $R_K$ is factorial. Regularity is obtained in the previous result (of course $R_K$ is not local, so regularity by itself does not imply factoriality.∎

There is an alternate proof which actually constructs a suitable $P$ so that $R_K$ is a localization of $R_P$ and for which $R_P$ is already factorial (explicit generators for the minimal primes can be found), but this is very tedious, and uses the notion of weak integral simplicity.

In Appendix A, there are some results about the Picard group (analogous to the class group for Dedekind domains) of rings of the form $R_P$ for $P$ satisfying the hypotheses of VI.6. This measures how far away a specific $R_P$ is from being itself factorial. In particular, if $P$ is irreducible (over $\mathbf{R}[x_i^{\pm 1}]$) and $R_P$ is regular, then $\mathrm{Pic}(R_P) = \mathbf{Z}^{e-(d+1)} \oplus \mathbf{Z}/f\mathbf{Z}$ where $d = \dim \mathrm{cvx}\,\mathrm{Log}\,P$, $e$ is the number of facets of K, and $f$ is some integer strictly greater than zero.

There are still a few loose ends to straighten out, in order to complete the proof of VI.1:

VI.8 **LEMMA** (Ad (i) implies (ii')). If K is a projectively faithful integral polytope in $\mathbf{R}^d$ such that $R_K$ is factorial,then K and dK satisfy the unique dominant stratum property.

Proof: By III.8A, K is locally solid, from which it follows that K is totally faithful, so by III.10, $R_{dK}$ is integral over $R_K$ and thus $R_{dK} = R_K$. Thus (i) implies (iii) applies for dK.∎

Obviously, (ii') implies (iii') follows from (ii) implies (iii) applied to dK.

Finally, if K is locally solid, we wish to show that K satisfying (ii) (respectively (iii)) implies that dK does as well. As in the argument in VI.8, $R_K = R_{dK}$. The minimal prime ideal $\mathbf{p}$ generated by $\{x^w/P \mid w \in (K \cap \mathbf{Z}^d)\backslash F\}$ (where Log P = $K \cap \mathbf{Z}^d$) in $R_K$ also generates the corresponding minimal prime (order) ideal in $R_{dK}$. By [H2; §4], there is a natural bijection between the pure states of $\mathbf{p}$ whose kernel has codimension 1, and the dominant strata of S = $(K \cap \mathbf{Z}^d)\backslash F$ with respect to some vertex (necessarily in the corresponding F). For each v in $d_eF$, there exists at least one pure state of $\mathbf{p}$ corresponding to it [H2; §III], so that the unique dominant stratum property is equivalent to $\mathbf{p}$ admitting exactly $\#(d_eF)$ of such pure states for each facet F. This is clearly independent of whether K or dK is chosen, so (iii) implies (ii') if K is locally solid.

To prove (iii) implies (iii') (when K is locally solid), it clearly suffices to show that (iii) implies (ii), for then (iii) ⇒ (ii) ⇒ (ii') ⇒ (iii').

So let v be a vertex of a facet F of K, and v' be the nearest lattice point to v on an edge L with $L \cap F = \{v\}$. Let $\alpha$ be a linear functional such that $\alpha|F = r$, $\alpha(v') = r+s$ $(s > 0$; and $\alpha|K \geq r)$, and thus $\alpha(w) \geq r+s$ for all w in $(K \cap \mathbf{Z}^d)\backslash F$. Say z in $(K \cap \mathbf{Z}^d)\backslash F$ yields a singleton dominant stratum with respect to v, distinct from v'. Let $\beta_i$ be linear functionals defining L, that is, $L = \cap\, \beta_i^{-1}(k_i) \cap K$ and $\beta_i|K \geq k_i$.

We have $\alpha(v') \leq \alpha(z)$ and $k_i = \beta_i(v') \leq \beta_i(z)$. Set $v'' = z+v-v'$. Then $\alpha(v'') \geq \alpha(v')$, and we obtain

$$\alpha(z) > \alpha(v'') > \alpha(v) \quad \text{and} \quad \beta_i(z) = \beta_i(v') \geq \beta_i(v).$$

Thus v" is in the cone generated by the $\beta_i$'s and $\alpha$. As K is locally solid, v" lies in $m(K \cap \mathbf{Z}^d)$ for sufficiently large m. Assume $v = \mathbf{0}$; we may write v" as a sum of m elements of $K \cap \mathbf{Z}^d$, not all of which are in F; select one of the terms $w_o$ in $K \cap \mathbf{Z}^d$ in this expression for v" that is not in F. Then $\alpha(z) > \alpha(w_o) > 0, \alpha(v)$, contradicting the hypothesis.∎

This completes the proof of the theorem. It is plausible to conjecture that when $R_K$ is factorial, then it is a localization of a pure polynomial algebra. One would have to be very careful about the generators. It is true however that if K is integrally simple, and $R_P$ is regular (where K = cvx Log P), then every localization of $R_P$ (and thus of $R_K$) at a prime ideal is a localization of a pure polynomial algebra. This follows easily from the definition of integrally simple, and our knowledge of localizations of $R_P$.

We shall see in section VIII, that if K is a projectively faithful simplex (that is an integral polytope), and K is integrally simple, then K is AGL(d,Z)–equivalent to n times the standard solid simplex for some integer n. So a rather crude property of $R_K$ serves to single out the nicest class of integral polytopes.

## VII. MEET-IRREDUCIBLE IDEALS IN $R_K$

Throughout this section, $K$ is an integrally simple polytope. We shall characterize all of the order ideals of $R_K$ that are meet-irreducible (as order ideals). This is equivalent to finding all of the the primitive ideals in $A^{\mathbf{T}}$, the fixed point C* algebra under the xerox type action of the d-torus $\mathbf{T}$, where the action is obtained from a character $P$ such that cvx Log $P = K$ and Log $P$ contains all the lattice points in each peak polytope of $K$ (see VI.6 and Appendices A and B).

Let us consider first what happens when $K$ is a simple as possible, i.e., the standard solid d-simplex. Set $P = 1 + \sum x_i$; upon setting $X_j = x_j/(1 + \sum x_i)$ for $j = 1,2,...,d$, as we have seen previously, $R_P = \mathbf{R}[X_i]$, with positive cone generated by $S = \{X_1, ..., X_d, X_{d+1} = 1 - \sum X_i\}$. The order ideals of either $R_P$ or $R_K$ are generated (as ideals in the respective ring) by products of elements of $S$. We can easily determine those ideals of this form that also happen to be meet-irreducible, either as order ideals or as ideals (in this case, probably in general, the two notions of meet-irreducibility coincide). We first observe that if $X_i X_j$ belongs to the meet-irreducible order ideal $I$ (with $i \neq j$), then either $X_i$ or $X_j$ belongs to $I$. Indeed, it is not very difficult to show that first, each $I_k$ defined as $I + X_k R_K$ is an order ideal (it is generated by $X_k$ and the generators of $I$), and second, they satisfy $I_i \cap I_j = I$. Meet-irreducibility thus would imply that either $X_i$ or $X_j$ belongs to I.

Similar considerations apply to $X_i^a X_i^b$ for positive integers $a$ and $b$—that is, if $X_i^a X_i^b$ belongs to I, then one of $X_i^a, X_j^b$ must also belong. It follows that $I$ is generated as an ideal by terms of the form $\{X_i^{a(i)}\}$ where $i$ runs over some subset of $\{1, 2, ..., d+1\}$, and $a(i)$ takes on positive integer values. Moreover, not every member of $\{1, 2, ..., d+1\}$ can be represented in the index set, else the ideal would be improper! It is straightforward to check in the pure polynomial algebra $R_P$, that all such ideals are meet-irreducible even with respect to the class of all ideals, not just order ideals, and the corresponding result thus also holds in $R_K$.

If instead we consider the standard hypercube, then a similar phenomenon occurs. Set $P = \prod(1+x_i)$; we have seen previously (section I, example B) that $R_P$ is the pure polynomial algebra in the variables $X_i = x_i/(1+x_i)$, and the positive cone is generated by $\{X_i, 1-X_i | i = 1, 2, ... d\}$. The meet-irreducible order-ideals (which again turn out to be meet-irreducible as ideals) are precisely those with generating set $\{Z_i^{a(i)}\}$, where each $Z_i$ is one of $\{X_i, 1-X_i\}$, the $i$ running over a subset (possibly the whole set) of $\{1, 2,..., d\}$. The theorem we ultimately prove (VII.11) about generators for the meet-irreducible order ideals of general $R_K$ with $K$ integrally simple, will clearly extend this type of characterization.

Let $F_1,...,F_e$ be a listing of the facets. The minimal prime ideals, $\mathfrak{p}_i$ of $R_K$ are generated by $\{x^w/P | w \in (K \cap \mathbf{Z}^d) \setminus F\}$ (take for $P$, any positive polynomial with Log $P = K \cap \mathbf{Z}^d$. By VI.1, each $\mathfrak{p}_i$ is principal, say with generator $z_i$; by V.1, we may take $z_i$ to be a positive element $R_P$, and an order unit of $\mathfrak{p}_i$. In fact, by V.I, we could choose

$$z_i = \sum_{w \in (K \cap Z^d) \setminus F_i} \frac{x^w}{P}.$$

Once the choice for the $z_i$ has been made, for $w = (w(1), \ldots, w(e))$ in $\mathbf{Z}^e$ denote by $z^w$, the product $z_1^{w(1)} z_2^{w(2)} \ldots z_e^{w(e)}$. We shall show that every order ideal that is meet-irreducible is of the form $z_1^{a(1)} R_K + z_2^{a(2)} R_K + \ldots + z_e^{a(e)} R_K$, with $a(i) \in \{0,1,2,\ldots\} \cup \{\infty\}$ (by convention, $z_i^\infty = 0$), subject to

$$\bigcap_{\{j \mid a(j) \neq \infty\}} F_j \neq \varnothing \qquad \text{and} \qquad a(j) \neq 0 \text{ for all } j.$$

Moreover, all such ideals are meet-irreducible, and it even follows that they are irreducible as ideals. This means that the generators for the primitive ideals of $A^T$ can be found explicitly.

We first require a result about cancellation of the ordering, analogous to that in II.1:

VII.1 **PROPOSITION** If $R_K$ is a unique factorization domain, and $z$ is a positive element of $R_K$ such that $za \geq 0$ for $a$ in $R_K$, then $a$ is positive in $R_K$.

Proof: As $z \geq 0$, there exist positive integers $N$ and $k$, and $w$ in $(kK) \cap Z^d$ such that $Nz \geq x^w/P^k$, where $P = \sum_{K \cap Z^d} x^v$. As $zR_K$ is an order ideal, there exists $s$ in $R_K$ such that $zs = x^w/P^k$. It suffices to show $s \geq 0$, for if so, then we write (after multiplying $a$ by an order unit, which does not affect $a$ being positive by II.5) $a = f/P^m$, for some $f$ with $\operatorname{Log} f \subset (kK) \cap Z^d$; we obtain $x^w f/P^m \geq 0$. Thus there exists M so that $x^w f P^m$ has only positive coefficients, hence the same is true of $fP^M$, so that $a \in R_K^+$.

Set $r = x^w/P^k$; then $rR_K$ is an order ideal. We may factor $r = a_1 a_2 \ldots a_j u$ with the $a_i$'s being irreducible elements of $R_K$ and $u$ a unit thereof. As $R_K$ is factorial, each $a_i R_K$ is a prime (and necessarily minimal prime) ideal. By V.4, each $a_i R_K$ is itself an order ideal. By V.1, for each i, either $a_i$ is positive or $-a_i$ is positive. As $u$ is a unit (and the pure state space of $R_K$ is connected), one of $\pm u$ is positive. If we replace $a_i$ by $-a_i$ whenever $a_i$ is negative, and adjust $u$ (by possibly multiplying by $-1$) so that the same equation still holds, then from $r$ being positive, we deduce that $u$ is positive. From $zs = a_1 a_2 \ldots a_j u$, we deduce that $z$ is a product of a subset of the $a_i$'s (notice that some of them may be equal—this is permissible), with a unit; we deduce that $s$ is a product of the remaining $a_i$'s with a unit. The unit, however, must be positive (rather than negative) as a result of $zs \geq 0$ and $z \geq 0$. Thus s is a product of positive elements, so is positive.∎

While not directly related to the rest of this section, the following weird result is a consequence of the preeceding:

**COROLLARY** Let $P$ in $\mathbf{R}[x_i^{\pm 1}]^+$ be such that $K = \text{cvx Log } P$ is integrally simple, and for all vertices $v$ of $K$, the lattice points of the peak polytope $K_v$ all belong to Log P. Let $F$ be a facet of $K$ and set $K' = \text{cvx}\{(K \cap \mathbf{Z}^d) \setminus F\}$. Let $Q$ in $\mathbf{R}[x_i^{\pm 1}]^+$ be such that $d_e K' \subset \text{Log } Q \subset K'$. If $f$ in $\mathbf{R}[x_i^{\pm 1}]$ has the property that there exists a positive integer $N$ with $Q^N f$ having only positive coefficients, then there exists $n$ so that $P^N f$ has only positive coefficients.

> **REMARK:** In other words, by performing a truncation on $K$ which amounts to removing one "layer", we obtain an implication about positivity; this is obviously related to II.1, but does not hold for more general polytopes than integrally simple ones.

Proof: As $K$ is locally solid, we may assume $\text{Log } f \subset \text{Log } P$ (replacing $P$ by a power of itself multiplied by a monomial, if necessary). Set

$$z = \sum_{K' \cap Z^d} \frac{x^w}{P} \in R_{K'},$$

$$a = f/P \in R_K, \text{ and}$$

$$Q' = \sum_{K' \cap Z^d} x^w = Pz \in \mathbf{R}[x_i^{\pm 1}].$$

Since $d_e \text{Log } Q' \subset \text{Log } Q \subset \text{Log } Q'$, there exists an integer $M$ such that $(Q')^M f$ has no negative coefficients (II.1); thus $z^M a$ belongs to $R_K^+$. However, $z$ is an order unit for the ideal of $R_K$ generated (as an ideal) by $\{x^w/P \mid w \in K' \cap \text{Log } P\}$, which is the minimal prime order ideal corresponding to the facet $F$. As order units in $R_K$ are invertible, by V.1, this ideal is just $zR_K$. We have $a z^M \geq 0$, so iterating the previous result, we obtain $a \in R_K^+$. As $a \in R_P$, we conclude that $a \in R_P^+$ by II.5.∎

VII.IA **LEMMA** Every (proper) order ideal of $R_K$ is generated as an ideal by a finite set of the form $\{z^v\}$ for some $v$'s in $(\mathbf{Z}^+)^e$.

Proof: Set $r = x^w/P^k$ with $w \in k\,\text{Log } P$ ($P$ as defined above). Then $rR_K$ is an order ideal; as the only minimal prime ideals that contain $r$ are order ideals (V.4), we obtain $v$ in $(\mathbf{Z}^+)^e$ such that $r = z^v u$, with $u$ a unit in $R_K$; as $r \geq 0$, $u$ is positive at one state, thus positive at them all, and so is an order unit. Hence $rR_K = z'R_K$; since every order ideal is generated as an ideal by terms of the form $x^w/P^k$, we are done.∎

**VII.2 LEMMA** All ideals of $R_K$ of the form

$$z^{w_1}R_K + z^{w_2}R_K + \ldots$$

(for $w_i$ in $\mathbf{Z}^e$) are order ideals.

Proof: Each $z_i$ generates an order ideal, as an ideal in $R_P$, hence in $R_K$; and products and sums of order ideals are order ideals.∎

**VII.3 LEMMA** If $Q$ is an order ideal of $R_K$ and $k(i) \in \mathbf{N} \cup \{\infty\}$ for $i = 1,2,\ldots,e$, then

$$\bigcap_i \{Q + z_i^{k(i)} R_K\} = Q + z^k R_K.$$

(Recall that $z^k = z_1^{k(1)} z_2^{k(2)} \ldots z_e^{k(e)}$.)

Proof: First note that both sides in the alleged equality are order ideals, and the left side contains the right. If any of the $k(i)$'s equal $\infty$, then equality is trivial, so we may assume that $k \in \mathbf{Z}^e$. Let $t = \#\{j \mid k(j) \neq 0\}$. If $t$ is 1 (or zero), then again equality is trivial. Re-index so that $k(j) > 0$ if and only if $j \leq t$.

Select $r$ in the positive cone of the left side; by interpolation, we may write

$$r = q + z_t^{k(t)} y \quad \text{in} \quad Q + z_t^{k(t)} R_K,$$

with $q$ in $Q^+$, and $z_t^{k(t)} y \geq 0$; by VII.1, $y$ itself is in $R_K^+$. By induction,

$$\bigcap_{i<t} \{Q + z_i^{k(i)} R_K\} = Q + \left(\prod_{i<t} z_i^{k(i)}\right) R_K;$$

as $r$ belongs to the left side of this, we have that

$$z_t^{k(t)} y = q' + \left(\prod_{i<t} z_i^{k(i)}\right) y'$$

with $q'$ in $Q^+$, and as before, $y'$ in $R_K^+$. Therefore,

$$\left(\prod_{i<t} z_i^{k(i)}\right) y' \leq N z_t^{k(t)} y$$

for some positive integer $N$. The left side of this expression therefore belongs to the order ideal, hence to the ideal generated by $z_t^{k(t)}$. We may thus write the left side as $z_t^{k(t)} y''$ for some $y''$ in $R_K$. By unique factorization (each $z_i$ is an irreducible), $y'$ must belong to $z_t^{k(t)} R_K$, say $y'' = z_t^{k(t)} y$, so $r = q + q' + z^k y'''$, as desired.∎

**VII.4 PROPOSITION** If $I$ is a meet-irreducible order ideal of $R_K$, then

$$I = z_1^{a(1)} R_K + z_2^{a(2)} R_K + \ldots + z_e^{a(e)} R_K,$$

for some selection of a(i)'s in $\{0,1,2,...\} \cup \{\infty\}$.

Proof: We know $I = z^{w1}R_K + z^{w2}R_K + ...$ for some selection of $w_i$ in $(\mathbf{Z}^+)^e$. Suppose that $z^w$ is an element of I; by the preceding result

$$\bigcap \{I + z_j^{w(j)}R_K\} = I + z^w R_K = I$$

(recall that $w = (w(1),...,w(e)) \in \mathbf{Z}^e$); as I is meet-irreducible, at least one of the $z_j^{w(j)}$ belong to it, and so $z^w$ may be replaced by $z_j^{w(j)}$ in the generating set, and we are done.∎

The next several lemmas are true far more generally:

VII.5 **LEMMA** If C is an order ideal in $R_K$ and z in $R_K^+$ generates an order ideal as an ideal (i.e., $zR_K$ is an order ideal), then $z^{-1}C = \{r \in R_K | zr \in C\}$ is an order ideal.

Proof: Suppose $0 \le a \le b$ with a in $R_K$, and b in $z^{-1}C$. Then $0 \le za \le zb \in C$, so $za \in C$, and thus a belongs to $z^{-1}C$. Hence $z^{-1}C$ is convex. Select x in $z^{-1}C$. Then $zx \in C \cap zR_K$. The latter is an order ideal, so we may find $y_1,y_2$ in $C \cap zR_K$ with $zx = y_1 - y_2$. Writing $y_i = zv_i$, we have that $v_i \ge 0$, so $x = v_1 - v_2$, and the $v_i$ belong to $(z^{-1}C)^+$. Hence $z^{-1}C$ is directed.∎

VII.6 **LEMMA** (a) Let $I_1, ..., I_m$, I be order ideals in $R_K$. Then

$$\left(\sum I_j\right) \cap I = \sum \left(I_j \cap I\right).$$

(b) If additionally $zR_K$ is an order ideal and z is positive, then

$$z^{-1}\left(\sum I_j\right) = \sum z^{-1}(I_j),$$

and these are order ideals.

Proof: (a) This follows from the fact that order ideals in $R_K$ are ideals, and from the modular law for ideals.

(b) This is completely routine.∎

VII.7 **LEMMA** Let I be the ideal generated by $\{z_i^{a(i)} | i = 1,2,...,e\}$. If $a(1) > 1$, then

$$z_1^{-1}(I) = z_1^{a(1)-1}R_K + \sum_{i=2}^{e} z_i^{a(i)}R_K.$$

Proof: We have that $z_1^{-1}I = \sum z_1^{-1}(z_i^{a(i)}R_K)$. For $i = 1$, it is clear that $z_1^{-1}(z_1^{a(1)}R_K) = z_1^{a(1)-1}R_K$. For $i > 1$, suppose that $z_1r$ belongs to $z_i^{a(i)}R_K$ (with i held fixed); that is $z_1r =$

$z_i^{a(i)}r'$ for some $r'$ in $R_K$. By unique factorization (and the fact that $i \neq 1$), $z_i^{a(i)}$ divides $r$, so $r$ belongs to the ideal it generates. Thus $z_1^{-1}(z_i^{a(i)}R_K) \subset z_i^{a(i)}R_K$ for $i \neq 1$, which yields the desired result.∎

VII.8 **LEMMA** If I, J, K are order ideals, then $I^{-1}(J \cap L) = I^{-1}J \cap I^{-1}L$ (we are using $I^{-1}J$ to mean $\{r \in R \mid rI \in J\}$; this differs somewhat from the notion used in Appendix A).

Proof: Again, this is routine.∎

VII.9 **LEMMA** If I is the order ideal generated by $\{z_i^{a(i)} \mid i = 1,2,\dots,e\}$, and it is not meet-irreducible, and if $a(1) > 1$, then the ideal I', generated by $\{z_i^{a(i)} \mid i = 2,\dots,e\} \cup \{z_1^{a(1)-1}\}$, is also not meet-irreducible.

Proof: Write $I = J \cap L$, where J and L are order ideals. Then

$$I' = z_1^{-1}(I) = z_1^{-1}(J \cap L) = z_1^{-1}(J) \cap z_1^{-1}(L),$$

and as $z_1^{-1}J$, $z_1^{-1}L$ are both order ideals, either I' is not meet-irreducible or one of $z_1^{-1}J$, $z_1^{-1}L$ is equal to $z_1^{-1}I$, which latter entails either J or L equalling I.∎

VII.10 **PROPOSITION** If I is the ideal generated by $\{z_i^{a(i)}\}$, and I is proper, then it is meet-irreducible.

Proof: If not, then by the above, the ideal I', generated by $\{z_i \mid a(i) \neq \infty\}$, would not be meet-irreducible; but I' is actually a prime ideal corresponding to the face, $\cap F_i$.∎

Combining the two most recent propositions, we have proved the following:

VII.11 **THEOREM** If $K$ is an integrally simple integral polytope, then the meet-irreducible order ideals of $R_K$ are precisely those obtained from the following prescription:

Define $P = \sum x^w$, the sum over $K \cap Z^d$. Begin with a face $F$, of dimension $m < d = \dim K$. Let $F_1, \ldots, F_{d-m}$ be a set of facets with $F = \cap F_j$. For each j define

$$z_j = \sum_{w \in Z^d \cap F_j} \frac{x^w}{P} \in R_K.$$

Let $a(1),\ldots,a(d-m)$ be positive integers, and set $I$ to be the ideal generated by

$$\{z_i^{a(i)} \mid i = 1, 2, \ldots, d-m\}.$$

Then $I$ is meet-irreducible, its associated prime ideal is $\mathfrak{p}_F$, generated by $\{z_1, \ldots, z_{d-m}\}$. Finally these choices for $I$ exhaust all the meet-irreducible ideals with $\mathfrak{p}_F$ as associated prime ideal.

## VIII. ISOMORPHISMS

In this section, our principal result is that for integral polytopes, K and K", any ring isomorphism between $R_K$ and $R_{K''}$ is automatically an order-isomorphism, provided either one is integrally simple, or is indecomposable in a rather weak sense. The latter condition is more or less generic in the sense that a random (finite) set of lattice points will generate an indecomposable integral polytope almost all of the time. At the level of $R_P$, this property fails; ring isomorphisms need not be even order-preserving. We also consider a group of automorphisms of $R_K$ which is conjectured to be of finite index in the group of all ring automorphisms.

In addition, we examine in some detail the nature of order-isomorphism of rings of the form $R_P$. We show that if P is irreducible as an element of $\mathbf{R}[x_i^{\pm 1}]$, then there are only finitely many order-isomorphisms and they are all induced by certain elements of AGL(d,**Z**) acting in the usual fashion. This contrasts with the situation in the case of $R_K$, where it is shown that there is a copy of $(\mathbf{R}^d)^{++}$ inside the group of order-automorphisms.

We first require a few results related to those in the proof of the main result of section I. As usual, K will denote an integral polytope.

VIII.1 **LEMMA** Let I be an order ideal of $R_K$, let L be a primary ideal of $R_K$ containing I, and let J denote the sum of all the order ideals inside L. Then J is an order ideal, and it is primary as an ideal.

<u>Proof:</u> Since finite sums and unions of ascending chains of order ideals are order ideals in rings having interpolation, J is an order ideal. By II.4, to show that J is primary, it is sufficient to show it is order-primary, that is, given a and b in $R_K^+$ such that $ab \in J$ and neither a nor b belongs to J, we show that for some integer n, $a^n$ and $b^n$ both belong to J. As $a, b \geq 0$, there exist integers m and M, and $w \in m(K \cap \mathbf{Z}^d)$, $v \in M(K \cap \mathbf{Z}^d)$ such that if $s = x^w/P^m$ (as usual, $P = \sum_{K \cap \mathbf{Z}^d} x^v$), $t = x^u/P^M$, then there exists a positive integer N so that $s \leq Na$ and $t \leq Nb$, with neither s nor t in J. Of course, $st \leq N^2 ab$ implies $st$ belongs to J.

Now, by II.2A, $sR_K$ and $tR_K$ are order ideals, so that $J + sR_K$ and $J + tR_K$ are order ideals. Hence s (say) belonging to L would imply that s belongs to J; therefore, neither s nor t belongs to L.

As $st$ belongs to L and L is primary, there exists n so that both $s^n$ and $t^n$ belong to L, and as in the last sentence of the preceding paragraph, this entails that they belong to J.∎

VIII.2 **COROLLARY** If $I$ is an order ideal of $R_K$, then all of its associated prime ideals are order ideals.

Proof: We may write [ZS; p.209, Theorem 4] $I = \cap \mathfrak{p}_i$ as an intersection of primary ideals; if $I \subset J_i \subset \mathfrak{p}_i$, where $J_i$ is the largest order ideal contained in $\mathfrak{p}_i$, as in VIII.1, then we have that the $J_i$ are primary and $I = \cap J_i$. If $\mathfrak{q}_i$ is the prime ideal associated to $J_i$, by V.4, it must be an order ideal, being a minimal prime ideal over $J_i$.∎

VIII.3 **LEMMA** If $K$ is locally solid, and $I$ is an order ideal that is principal as an ideal (e.g., $I = (x^w/P^m)R_K$ for $w$ in $m(K \cap \mathbf{Z}^d)$), then all of its associated prime ideals are the minimal primes corresponding to certain facets of K.

Proof: By [ZS; Theorem 14, p. 277], and the fact that $R_K$ is integrally closed (III.2), all associated primes over principal ideals are minimal. They are order ideals by VIII.2, so they are given by facets of K, as in [H1; §VII].∎

VIII.4 **LEMMA** If $K$ is locally solid, $w \in k(K \cap \mathbf{Z}^d)$, and $I \subset \cap \mathfrak{p}_F$ where the $F$ vary over all the facets of $K$ such that $w \notin kF$, t hen there exists an integer $n$ such $I^n \subset (x^w/P^k)R_K$.

Proof: As $(x^w/P^k)R_K$ is an order ideal, it follows from the previous results that its associated primes are of the form $\mathfrak{p}_F$ for $F$ facet. If $w$ does not belong to $kF$, then $x^w/P^k$ belongs to $\mathfrak{p}_F$, otherwise, $x^w/P^k$ does not belong to $\mathfrak{p}_F$. Hence the radical of $(x^w/P^k)R_K$ is exactly $I$, and as $R_K$ is noetherian, there exists $n$ so that $I^n \subset (x^w/P^k)R_K$.∎

Over the next several lemmas, we take a closer look at what happens to the vertices when two integral polytopes are added together:

VIII.5 **LEMMA** If $K_1$ and $K_2$ are two compact convex polytopes in $\mathbf{R}^d$, and $K = K_1 + K_2$ is their sum, then each vertex $v$ of $K$ is uniquely expressible as $v = w_1 + w_2$, with $w_i$ in $K_i$, and this entails that $w_i$ belong to $d_eK_i$.

Proof: There exist $w_i$ in $K_i$ so that $v = w_1 + w_2$. If $w_i = \alpha u_1 + (1-\alpha)u_2$ with $u_i$ in $K_i$, and $\alpha$ a real number in the interval $(0,1)$, we would obtain $v = \alpha(u_1 + w_2) + (1-\alpha)(u_2 + w_2)$, contradicting $v$ being a vertex. Hence $w_i \in d_eK_i$.

If $v = w_1 + w_2 = w_1' + w_2'$, we see that $v_i = \frac{1}{2}(w_i + w_i')$ belongs to $K_i$ but not to $d_eK_i$ for at least one of $i = 1,2$. If $w_1 \neq w_1'$, we obtain $v = v_1 + v_2$ contradicting the result of the previous paragraph. Hence $w_i = w_i'$, so uniqueness is proved.∎

**VIII.6 LEMMA** Let $f_1, f_2$ be elements of $\mathbf{R}[x_i^{\pm 1}]$. Then

$$\text{cvx Log } f_1 f_2 = \text{cvx Log } f_1 + \text{cvx Log } f_2.$$

Proof: One inclusion is clear. Let $v$ be a vertex of the right hand side. There exist $w_i$ in $d_e \text{cvx Log} f_i$ such that $v = w_1 + w_2$. By the uniqueness result above, $x^v$ has coefficient not zero in $f_1 f_2$. Hence $v$ belongs to $\text{Log } f_1 f_2$, and the other inclusion follows.∎

**VIII.7 PROPOSITION** Let $K$ be a locally solid integral polytope. Suppose $s$ is an element of $R_K$ such that there exists $t$ in $R_K$ with $st = x^w/P$ for some $P$ in $\mathbf{R}[x_i^{\pm 1}]^+$ having $\text{cvx Log } P = K$ and with $w \in K \cap \mathbf{Z}^d$. Then one of $\pm s$ is positive in $R_K$.

Proof: By translation, we may assume that $w = \mathbf{0}$, so $st = 1/P$. We may write $s = f/Q$, $t = g/Q'$ where $f, g, Q, Q'$ are in $\mathbf{R}[x_i^{\pm 1}]$, with $Q$ and $Q'$ having only positive coefficients, $\text{Log } f \subset \text{cvx Log } Q$, $\text{Log } g \subset \text{cvx Log } Q'$, and $\text{cvx Log } Q = eK$, $\text{cvx Log } Q' = e'K$ for integers $e$ and $e'$. We deduce $fgP = QQ'$.

Set $\underline{f} = \sum_{\text{Log } f} x^w$ and $\underline{\underline{g}} = \sum_{\text{Log } g} x^w$. By VIII.6, $\text{cvx Log } f\underline{\underline{g}} = \text{cvx Log } fg$ (as both equal $\text{cvx Log } f + \text{cvx Log } g$), and thus $\underline{f}\underline{\underline{g}}P/QQ'$ belongs to $R_K$ (as $K$ is locally solid). Therefore, $\underline{f}\underline{\underline{g}}P/P^{e+e'}$ belongs to $R_K$, and this is $a = \underline{f}\underline{\underline{g}}/P^{e+e'-1}$.

Set $\lambda_1 = \min\{|\text{nonzero coefficients of } f|\}$, and $\lambda_2$ will be the corresponding real number for $g$. Then $\underline{f} \geq (1/\lambda_1) f$, and $\underline{\underline{g}} \geq (1/\lambda_2) g$, and thus $\underline{f}\underline{\underline{g}} \geq (1/\lambda_1 \lambda_2) fg$. Thus, as $fgP = QQ'$, we deduce that $\underline{f}\underline{\underline{g}}P \geq (1/\lambda_1 \lambda_2) QQ'$. Hence $\underline{f}\underline{\underline{g}}P/QQ'$ is an order unit of $R_K$, so is invertible.

Also, $\text{cvx Log } \underline{f}\underline{\underline{g}}P = eK$, and $\text{Log } f = \text{Log } \underline{f}$ (similarly with $g$), so $f\underline{\underline{g}}P/\underline{f}\underline{\underline{g}}P = f/\underline{f} \in R_K$ and by the same means we deduce $g/\underline{\underline{g}}$ also belongs to $R_K$. Now

$$\frac{f}{\underline{f}} \frac{g}{\underline{\underline{g}}} \frac{P}{P} = \frac{QQ'}{\underline{f}\underline{\underline{g}}P} = \frac{fgP}{\underline{f}\underline{\underline{g}}P}.$$

As the last term is an order unit of $R_K$, we deduce that $f/\underline{f}$, $g/\underline{\underline{g}}$ are also invertible, and therefore, up to sign are positive, in fact order units. By multiplying $f$ and $g$ by $-1$ if necessary, we may assume that they are order units (as $fgP = QQ'$). Now

$$\frac{f}{Q} \frac{QQ'}{\underline{f}\underline{\underline{g}}P} = \frac{f}{\underline{f}} \frac{Q'}{\underline{\underline{g}}P}.$$

The latter term is positive (as $f/\underline{f}$ is an order unit), so as $(QQ')/\underline{f}\underline{\underline{g}}P$ is an order unit, we deduce that $s = f/Q$ is also positive.∎

**VIII.8 LEMMA** Let $K$ be a locally solid integral polytope, and let $P$ be an element of $\mathbf{R}[x_i^{\pm 1}]^+$ such that $\mathrm{cvx\,Log}\,P = K$. Suppose that $s$ is an element of $R_P$ and there exists $t$ in $R_K$ with $st = x^w/P^k$ for some $w$ in $k\,\mathrm{Log}\,P$. Suppose furthermore that $s = f/P^r$ with $d_e\mathrm{Log}\,f = d_e(eK)$ for some integer $r$. Then $s$ is an order unit of $R_P$.

Proof: There is a natural inclusion $R_P \subset R_K$. By the previous result, $s$ is in $R_K^+$. By the hypotheses on the vertices of $\mathrm{cvx\,Log}\,f$, $\gamma_v(s) \geq 0$ for all vertices $v$ of $K$ ($\gamma_v$ is the pure state corresponding to $v$ as in [H1;VII]). By [H1; V.4] applied to $R_{PQ}$ (where $\mathrm{Log}\,Q = K \cap \mathbf{Z}^d$), $s$ is an order unit of $R_{PQ}$, hence of $R_K$. Therefore $\beta(s)$ is strictly positive for all pure states $\beta$ of $R_K$. However, the construction of the pure states in [H1; §III] yields that the states of $R_P$ are the same as those of $R_K$ (even though $R_K$ need not be a localization of $R_P$), hence $s$ is an order unit in $R_P$.∎

**VIII.9 COROLLARY** Let $b, f$ be elements of $\mathbf{R}[x_i^{\pm 1}]$ be such that $bf = P \in \mathbf{R}[x_i^{\pm 1}]^+$. Suppose $Q$ in $\mathbf{R}[x_i^{\pm 1}]^+$ satisfies either:

$$\mathrm{Log}\,Q = (e\cdot\mathrm{cvx\,Log}\,P) \cap \mathbf{Z}^d \quad \text{for some integer } e \geq d,$$

or

$$\mathrm{Log}\,Q = (e\cdot\mathrm{cvx\,Log}\,P) \cap \mathbf{Z}^d \quad \text{for some integer } e \text{ and } e\cdot\mathrm{cvx\,Log}\,P \text{ is locally solid.}$$

Then there exists an integer $N$ so that $Q^N f$ has no negative coefficients.

Proof: Setting $K = \mathrm{cvx\,Log}\,Q = e\cdot\mathrm{cvxLog}\,P$, we see that in either case, $R_K$ is integrally closed. Without loss of generality, we may assume that $\mathbf{0} \in \mathrm{Log}\,b \subset \mathrm{Log}\,f \subset \mathrm{Log}\,P$. By VIII.6, $\mathrm{Log}\,f$ and $\mathrm{Log}\,b$ are each contained in $\mathrm{cvx\,Log}\,P$, so $f/Q$, $b/Q$ belong to $R_K$. Obviously $P^e/Q$ is an order unit of $R_K$, so is invertible in $R_K$. As

$$\frac{f}{Q}\frac{Q}{P^e}\frac{b}{Q}\frac{Q}{P^e} = \frac{1}{P^{2e-1}},$$

VIII.7 applies.∎

**REMARK:** In one sense at least, the result of VIII.9 is best possible. It is not true that $\mathrm{Log}\,Q = \mathrm{Log}\,P$ is sufficient to guarantee the same conclusion. For example, (with $d = 1$), set $P = x^3+x+2$, $f = x^2-x+2$, and $b = x + 1$. If $Q$ is any monic polynomial of degree $n$ with no $x^{n-1}$ term, then the coefficient of $x^{nN+1}$ in $Q^N f$ will always be $-1$.

Let $K, K'$ be two (real) polytopes in $\mathbf{R}^d$. They are homothetic if there exist a positive real number $\alpha$ and a point $r$ in $\mathbf{R}^d$ such that $K = \alpha K' + r$.

The polytope K is <u>indecomposable</u> if (for polytopes $K_i$) whenever $K = K_1 + K_2$, at least one of the $K_i$ is homothetic to K. If K and K' are rational and homothetic, then we may take the $\alpha$ and r to be rational; and a rational polytope that indecomposable as a rational polytope is indecomposable as a real one.

However, for integral polytopes, there is a different form of indecomposibility; we say the integral polytope K is <u>integrally indecomposable</u> if the $K_i$ are limited to integral polytopes (real homothety is used). For example, the quadrilateral K with vertices $\{(0,0), (1,0), (0,1), (2,2)\}$ is integrally indecomposable but 2K is not! We only occasionally refer to this notion of indecomposability, because of this unstable behaviour.

If p belongs to $\mathbf{R}[x_i^{\pm 1}]$ and cvx Log p is integrally indecomposable, then p is irreducible, that is, it does not factor non-trivially. Let K be an indecomposable integral polytope having $\mathbf{0}$ as a vertex. Let e be the greatest common divisor of all the coordinates of all the vertices of K; then $\breve{K} = (1/e)K$ is an integral polytope which we shall call the <u>reduced form</u> of K. Of course, $\breve{K}$ is indecomposable when K is; moreover, if p in $\mathbf{R}[x_i^{\pm 1}]$ has cvx Log p being in its reduced form, then p is necessarily an irreducible element of the Laurent polynomial algebra.

In general, an element c of a commutative domain A is called <u>irreducible</u> if $c = ab$ for a and b in A implies that one of a or b is a unit. If $P = \sum x^w$ (the sum over $\breve{K}$), and v belongs to Log P, then $x^v/P \in R_P \subset R_{\breve{K}} \subset R_K$ (note that $\breve{K}$ need not be projectively faithful or locally solid when K is, so we must be a little careful in this regard). We will show that these elements, $x^v/P$, are irreducible in $R_K$ (hence also in $R_P$).

(Cancellation) **LEMMA** Let L, V, V' be compact convex polyhedra inside $\mathbf{R}^d$.

$$\text{If } \quad L + V = L + V', \text{ then } V = V'.$$

<u>Proof</u>: Let v be a vertex of V; we first find a linear functional $\alpha$ and a point u of L such that

$$\alpha(v) > \alpha(z) \qquad \text{for all } z \in V\backslash\{v\};$$

$$\alpha(v+u) > \alpha(x) \qquad \text{for all } x \in (V+L)\backslash\{v+u\}.$$

There exists a linear functional $\alpha'$ exposing v as a vertex of V (that is, satisfying the first of the two conditions above). If $t = \max\{\alpha'(u) \mid u \in L\}$, let $L_o$ be the face of L given as $(\alpha')^{-1}(t) \cap L$. By slightly perturbing $\alpha'$, we may find $\alpha$ exposing v (as a vertex of V) and so that additionally, the face corresponding to $L_o$ is a singleton $\{u_o\}$. (This is because the set of functionals, for our purposes, is a space of the form $(\mathbf{R}^d\backslash\{0\})/\mathbf{R}^+$, which is homeomorphic to the sphere $S^{d-1}$; the set of those that expose a face of dimension exceeding 0 on L is contained in a finite union of lower dimensional subsets.) Clearly $\alpha(v+u_o) > \alpha(x)$ for all x in $(V+L)\backslash\{v+u_o\}$ so that $v+u_o$ is a vertex of $V + L$.

Now apply VIII.5 to $v+u_0$ as a vertex of $V'+L$; we may write $v+u_0 = v_1 + u_1$, where $v_1, u_1$ are vertices of $V'$, $L$ respectively. By the two inequalities above, it easily follows that $\alpha(u_1)$ attains the maximum value of $\alpha$ on L, so that $u_1 = u_0$. Thus $v = v_1$, and so $V'$ contains all the vertices of V, and thus $V \subset V'$; a symmetric argument yields the reverse inequality.∎

VIII.10 **LEMMA** If K is an indecomposable integral polytope, and P in $\mathbf{R}[x_i^{\pm 1}]^+$ is such that $\mathrm{cvx\,Log}\,P = K$, then for v in Log P, the element $x^v/P$ of $R_K$ is irreducible.

Proof: Suppose that $x^v/P = (f/Q)(g/Q')$ with f, g, Q, Q' in $\mathbf{R}[x_i^{\pm 1}]$, the latter two with positive coefficients, and $\mathrm{cvx\ Log}\ Q = mK$, $\mathrm{cvx\ Log}\ Q = m'K$ for positive integers m, m'. Then $x^v QQ' = fgP$. Without loss of generality we may assume that $v = \mathbf{0}$, so that $fgP = QQ'$. As $\mathrm{cvx\,Log}\,fg = \mathrm{cvx\ Log}\ f + \mathrm{cvx\ Log}\ g$, we have that

$$\bar{K} + \mathrm{cvx\ Log}\ f + \mathrm{cvx\ Log}\ g = (m+m')K = (m+m')e\bar{K}.$$

Now $\mathrm{cvx\ Log}\ f = me\bar{K}$, and $\mathrm{cvx\ Log}\ g = m'eK$, so we must have by the indecomposability of $(m+m')e\bar{K}$ that $\mathrm{cvx\ Log}\ f = a\bar{K} + v$, $\mathrm{cvx\ Log}\ g = a'\bar{K} + v'$. By the preceding lemma, $\mathrm{cvx\ Log}\ f + \mathrm{cvx\ Log}\ g = ((m+m')e - 1)\bar{K}$, so $v+v' = 0$ and $a + a' = (m+m')e-1$. Since $\mathrm{cvx\ Log}\ f = me\bar{K}$, we have $a \le me$ and similarly $a' \le m'e$. Since cvx Log f is an integral polytope, a and a' must be integers. Thus either $a = me$ or $a' = m'e$ holds. Either entails $v = v' = 0$. From $a = me$, we would deduce that f/Q satisfies the conditions of VIII.8 and so would be an order unit in $R_K$, and $a' = m'e$ would similarly result in g/Q' being an order unit.

In $R_{\bar{K}}$, all elements that become units in $R_K$ are already units in the smaller ring, so $x^v/P$ is still irreducible therein.∎

VIII.11 **LEMMA** Let K be a locally solid indecomposable integral polytope with reduced form $\bar{K}$. Let v be a vertex of $\bar{K}$. Set $P = \sum x^w$, the sum taken over $K \cap \mathbf{Z}^d$, and set $s = x^v/P$ in $R_K$. Then s satisfies the following:

(i) $\beta(s) > 0$ for all point evaluations $\beta$ (that is, there exists r in $(\mathbf{R}^d)^{++}$, so that $\beta$ sends s to s(r));

(ii) $\beta_v(s) = 1$ where $\beta_v$ is the unique pure state corresponding to the vertex v;

(iii) $\beta(s) = 0$ if v does not belong to the face of $\beta$, $F(\beta)$ (see section IV);

(iv) If s' in $R_K$ satisfies (i), (ii), (iii), then there exists n such that $(s')^n = sa$ for some a in $R_K$;

(v) s is irreducible;

(vi) s is unique, up to associates, with respect to properties (i )through (v).

Proof: Properties (i), (ii), (iii) are obvious, and (v) is proved in the previous lemma.

(iv). If s' satisfies (iii), then $s' \in \cap \mathbf{p}_F$, the F varying over the facets of K disjoint from v. By VIII.4, there exists n so that $(\mathbf{p}_F)^n \subset sR_K$, so $s'R_K \subset sR_K$.

(vi). Suppose s" satisfies ( i) through (v). By (iv), there exists n so that $s^n = s''a$ for some a in $R_K$. Write $s'' = f/Q$, $a = g/Q'$, with f, g, Q, Q' in $\mathbf{R}[x_i^{\pm 1}]$, the latter two with positive coefficients, and satisfying cvx Log Q = e$\breve{K}$, cvx Log Q' = e'$\breve{K}$, e and e' being integers. As usual, we may assume that $v = \mathbf{0}$ and $K \subset (\mathbf{R}^d)^+$. From $s^n = s''a$, we deduce $fgP^n = QQ'$; thus

$$\text{cvx} \operatorname{Log} f + \text{cvx} \operatorname{Log} g + n\breve{K} = (e+e')\breve{K}.$$

As K and $\breve{K}$ are indecomposable and contained in the positive orthant, we must have that cvx Log f = t$\breve{K}$ for some integer t. Thus $e \le t$. By VIII.8, $f/P^t$ is an order unit of $R_K$ and we obtain

$$s'' = \frac{1}{P^{t-e}} \frac{1}{P^t} \frac{P^e}{Q}.$$

As $P^e/Q$ is an order unit and s" is not, it must be that $e < t$; recalling our assumption that $v = \mathbf{0}$, we deduce $s'' \in (x^v/P)R_K$. As s" is irreducible, it is an associate of $s = x^v/P$.∎

VIII.12 **THEOREM** Let $K_1$ and $K_2$ be indecomposable integral polytopes with one of them being locally solid. Then any ring isomorphism $\phi: R_{K_1} \to R_{K_2}$ is an order isomorphism.

Proof: Rewrite $R_{K_i}$ as $R_i$. We may assume $K_i$ are both projectively faithful. As K being locally solid is equivalent to $R_K$ being integrally closed in its field of fractions, both $K_i$'s are locally solid. We first observe that $\phi$ carries maximal ideals of $R_2$ to $R_1$, via $\phi^*(M) = \phi^{-1}(M)$. These are precisely the kernels of pure states, so we see that $\phi$ induces a continuous map $\phi^*: d_eS(R_2,1) \to d_eS(R_2,1)$, which is obviously a homeomorphism.

We may suppose that both $K_1$ and $K_2$ are sitting inside $\mathbf{R}^d$, and that $K_1$ is d-dimensional. As the $K_i$ are homeomorphic to the pure state spaces of the $R_i$ (IV.I), we obtain that $K_2$ is also d-dimensional. Now the identification, in Section IV, of the $K_i$ with the pure state spaces, identifies the interior of $K_i$ with the pure states which have a neighbourhood that looks like a d-ball. Hence $\phi^*$ maps the boundaries homeomorphically.

Let F be a facet of $K_2$. We claim there exists a facet G of $K_1$ such that

$$\{\alpha \in d_eS(R_1,1) \mid F(\alpha) \subset G\} = \{\phi^*(\beta) \mid \beta \in d_eS(R_2,1) \text{ and } F(\beta) \subset F\}.$$

To this end, consider the minimal prime ideal $\mathfrak{p}_F$ of $R_2$. Then $\phi^{-1}\mathfrak{p}_F$ is a minimal prime ideal of $R_1$. Moreover, $E_F$, defined as $\{\beta \in d_eS(R_2,1) \mid F(\beta) \subset F\} = \{\beta \in d_eS(R_2,1) \mid \beta|\mathfrak{p}_F = 0\}$ is a d–1-dimensional set (homeomorphic to F via $\Lambda_P$, see section IV) inside the boundary. Therefore, its image under $\phi^*$ is also a d–1-dimensional subset of the boundary of $d_eS(R_1,1)$. As the boundary is a union of sets of the form (for G running over the facets of K):

$$D_G = \{\alpha \mid F(\alpha) \subset G\},$$

there exists a facet G of $K_1$ such that $D_G \cap \phi^*E_F$ is d–1-dimensional (that is, it contains a d–1-ball). Since we are dealing with rational functions, anything that vanishes on $D_G \cap \phi^*E_F$ must vanish on all of $D_G$. Thus $D_G \subset \phi^*E_F$, so that $\mathfrak{p}_G$ is a minimal prime ideal of $R_1$, corresponding to G, and it contains $\phi^{-1}\mathfrak{p}_F$. The former being a minimal prime ideal, we deduce that $\mathfrak{p}_G = \phi^{-1}\mathfrak{p}_F$, and it follows that $D_G = \phi^*E_F$. Thus $\phi$ induces a function on facets, via $\tilde{\phi}(F) = G$.

Now if $F_o$ is an arbitrary face of $K_2$, with say $F_o = \cap F^i$, the $F^i$ being facets, then

$$\mathfrak{p}_{F_o} = \sum \mathfrak{p}_{F^i}$$

(just check the definitions). Hence

$$\phi^{-1}(\mathfrak{p}_{F_o}) = \sum \phi^{-1}(\mathfrak{p}_{F^i})$$

so we may extend $\tilde{\phi}$ to a function on faces, via $\tilde{\phi}(F_o) = \cap\, \tilde{\phi}(F^i)$. In particular, if $v_o$ is a vertex of $K_2$, then $\tilde{\phi}(v_o)$ is a vertex of $K_1$. In other words, $\phi^*$ sends the pure states corresponding to vertices of $K_2$ to those corresponding to vertices of $K_1$.

Form the reduced integral polytopes, $\tilde{K}_1$ and $\tilde{K}_2$. Since we are only going to regard these as pure state spaces, the preceding applies with the reduced forms replacing the originals. Let v be a vertex of $\tilde{K}_2$, with corresponding pure state $\beta_v$ (with $F(\beta_v) = \{v\}$). Then $\phi^*\beta_v = \beta_{\tilde{\phi}v}$ (of course, we know now that $\tilde{\phi}(v)$ is a vertex of $K_1$) is a restatement of the result of the preceding paragraph. Set $s_v = x^v/P$, where $P = \sum x^w$, the sum being taken over $K \cap \mathbf{Z}^d$. Then $s_v$ in $R_2$ satisfies (i) through (v) of the preceding lemma, so that $\phi^{-1}(s_v)$ also does as an element of $R_1$. Thus if we define $s_{1,v}$ as $x^{\tilde{\phi}v}/P_1$ where $P_1 = \sum x^w$, the sum being taken over $K_1 \cap \mathbf{Z}^d$, then $\phi^{-1}(s_v)$ and $s_{1,v}$ are associates; say their ratio is u. As u is a unit, and both elements are positive, we must have that u is an order unit (V.1) ).

Set $Q = \sum x^w$, the sum taken over $K_2 \cap \mathbf{Z}^d$. For w in $K_2 \cap \mathbf{Z}^d$, we may write

$$Nw = \sum_{v \in d_e K_1} \lambda_v \tilde{\phi}(v) \qquad \text{for some positive integers } N, \lambda_v, \text{ and so that}$$

$$\left(\sum \lambda_v\right) K_1 = N K_1.$$

It follows that

$$\left(\frac{x^w}{Q}\right)^N = \prod_{v \in d_e K_1} \left(\frac{x^{\tilde{\phi}(v)}}{P}\right)^{\lambda_v} \cdot (\text{order unit}),$$

and moreover, up to multiplication by an order unit, $x^w/Q$ is the only solution to this equation which is non-negative at all pure states. Thus $x^w/Q$ is up to multiplication by an order unit, a product of terms of the form $\phi^{-1}(x^v/P)$, hence is in the image of $R_2^+$ under $\phi^{-1}$, and moreover $\phi^{-1}(R_2^+)$ is contained in $R_1^+$. Thus $\phi$ is an order-isomorphism.∎

Attempts to prove this type of result in the decomposable case along these lines founder, because of the following phenomenon. Suppose $nK = \sum K_i$, with the $K_i$ being indecomposable integral polytopes, and consider the polynomials $P_i = \sum x^w$, the sum over $w$ in $K_i \cap Z^d$. For $v$ in $d_e K_i$, the element $x^v/P_i$ ought to be irreducible (as its analogue is in the indecomposable case), and irreduciblity of this type of element was crucial to the proof above. However, in general it is not, even when $K$ is integrally simple (Example VIII.19).

Another situation where it is possible to prove ring isomorphism implies order-isomorphism (and with an easier proof) occurs when $K$ is integrally simple:

VIII.13 **PROPOSITION** Let $K_1$ be an integrally simple polytope and let $K_2$ be an integral polytope. Then any ring isomorphism $\phi: R_{K_1} \longrightarrow R_{K_2}$ is an order isomorphism.

<u>Proof</u>: We may assume that $K_2$ is projectively faithful. As $R_1$ is factorial, so must $R_2$ be, and thus $K_2$ is also integrally simple. As in the first half of the proof of VIII.12 (which does not involve the indecomposability of the polytopes), $\phi$ induces in a natural way a bijection $\tilde{\phi}$ between the faces of $K_2$ and those of $K_1$.

Let $F$ be a facet of $K_2$, and form the *principal* minimal ideal $\mathfrak{p}_F$; by V.1, there exists positive $z_F$ in $R_2^+$ such that $\mathfrak{p}_F = z_F R_2$. Viewed as a function on $d_e S(R_2,1)$, $z_F$ is positive, and vanishes only on the boundary, and then only at those pure states corresponding to $F$ (that is, for a pure state $\beta$, $\beta(z_F) = 0$ if and only if $F(\beta) \subset F$). Obviously, $\phi^{-1}(z_F) = y$ generates a minimal prime ideal, call it $\mathfrak{q}$ (we do yet know if $\mathfrak{q}$ is an order ideal). Now the zero set of $y$ on the pure state space of $K_1$ is precisely the complement of the set of pure states corresponding to

$\bar{\phi}F = G$, say. Define $y_G = \sum x^w/P$, where $P$ is any positive polynomial with $\operatorname{Log} P = K_1 \cap \mathbf{Z}^d$ and the sum is over those $w$ in $(K_1 \cap \mathbf{Z}^d) \setminus G$. By VI.1 and V.1, $y_G R_1$ is a minimal prime ideal, and obviously the maximal ideals containing this are the same as those containing $yR_1$. Now $y_G R_1 = \mathbf{p}_G$ is exactly the intersection of all the kernels of pure states corresponding to $G$ (the quotient is given as in §VII of [H1]), so that $y$ belongs to $y_G R_1$; by minimality, we must have that $yR_1 = y_G R_1$, and since the latter is an order ideal and $y$ is positive at at least one pure state, it follows from V.1, that $y$ is positive in $R_1$.

By VI.2, the $z_F$'s generate the positive cone of $R_2$, so that $\phi^{-1}$ is order-preserving. As the $y_G$'s (any choice, one for each facet G) generate the positive cone of $R_1$, we have that $\phi^{-1}$ maps $R_2^+$ onto $R_1^+$. So $\phi^{-1}$, and thus $\phi$, is an order-isomorphism.∎

Of course, VIII.12 and VIII.13 apply to very different classes of integral polytopes; for example, any cartesian product of integrally simple polytopes will be integrally simple, but will not be indecomposable.

On the other hand, VIII.12 and VIII.13 fail if ring isomorphism is replaced by endomorphism (and we ask merely that the endomorphism be order-preserving), and they also fail on the level of $R_P$. There exists an $R_P$—in fact, $P = 1+x+y$—with a ring automorphism that is not order-preserving. We shall briefly discuss some examples before considering the automorphisms of $R_K$.

VIII.14 **EXAMPLES** Let $K$ be the standard 2-simplex and $P = 1+x+y$. Set $X = x/(1+x+y)$ and $Y = y/(1+x+y)$, so that $R_P = \mathbf{R}[X,Y]$ with positive cone generated multiplicatively and additively by $\{X, Y, 1-X-Y\}$. The transformation $\Lambda_P$ (see section IV) restricted to a map $(\mathbf{R}^d)^{++} \longrightarrow \operatorname{Int} K$ simply sends $(x,y)$ to $(X = x/(1+x+y), Y = y/(1+x+y))$, so that the pure state space of $R_P$, when viewed as polynomials in $X$, $Y$, is $K$ itself, i.e., point evaluations at points of K. An endomorphism (as rings) of $R_P$ will extend to an endomorphism of $R_K$ precisely if it sends order units to order units. Since the order units of $R_P$ are exactly the elements that are strictly positive at all pure states, it is necessary and sufficient that the induced function on $\operatorname{Spec} R_P$ (which of course contains K) map K into itself. Any order-preserving endomorphism will do this.

Define $\phi_1: R_P \longrightarrow R_P$ via $\phi_1(X) = X$, $\phi_1(Y) = X+Y$. This clearly extends to an automorphism (as a ring) of $R_P$, and equally clearly does not leave $K$ invariant and thus is not order-preserving—in fact, $\phi(1-X-Y) = 1-2X-Y$, which is not positive. So we have a ring automorphism of $R_P$ that is not order-preserving.

Somewhat more subtle is the following example of an endomorphism of $R_K$ which is not order-preserving. Define $\phi_2: R_P \longrightarrow R_P$ via $\phi_2(X) = X$, $\phi_2(Y) = Y(X^2 + Y^2 - XY)$. As $(X,Y)$

range over the points of the unit triangle $(X \geq 0, Y \geq 0, 1 \geq X+Y)$, it is straightforward to see that so do their images under the map induced by $\phi_2$, $(\phi_2(X) = X, \phi_2(Y))$, so that $\phi_2$ extends to a ring endomorphism of $R_K$, which we also call $\phi_2$. Now $\phi_2$ is not order-preserving (which implies, by either VIII.12 or VIII.13, that it is not onto). In fact we claim that $\phi_2(Y)$ is not positive in $R_K$. By II.1, it is sufficient to show that it is not positive as an element of $R_P$, and by VII.1, it is sufficient to show that $X^2+Y^2-XY$ is not positive in $R_P$; c.f., [H4; §IV].

If the latter element were positive, then $X^2+Y^2 \geq XY$; thus $XY$ would belong to the order ideal generated by $X^2$, $Y^2$. However, this is just the ideal generated by $X^2$, $Y^2$ in $R_P = \mathbf{R}[X,Y]$. Since there are commutative algebras containing elements a, b such that $a^2 = b^2 = 0$ but $ab \neq 0$, $XY$ does not belong to the ideal generated by $X^2$, $Y^2$, and so $\phi(Y)$ is not positive.∎

In one variable, endomorphisms of $R_K$ *are* order-preserving.

Now we discuss automorphisms of $R_K$, at least for K being either integrally simple or indecomposable. Let $\tilde{\phi}$ be the map induced on K by the automorphism $\phi$. Then we know from the proofs of VIII.12 and VIII.13 that $\tilde{\phi}$ permutes the vertices, and knowing the values $\phi(x^v/P)$ as v varies over the vertices of K, determines $\phi$. Let us assume that $\tilde{\phi}$ acts trivially on the vertices (that is, $\phi$ is in the kernel of the representation as a permutation group on the vertices—we shall have more to say about the "geometric" transformations that act non-trivially on the vertices, when we discuss order automorphisms of certain $R_P$, later on in this section). Then it is straightforward to check that in fact, $\phi$ acts trivially on the set of faces of K. Let us describe in some more detail what $\phi$ must look like under these conditions.

We may assume that one of the vertices of K is $\mathbf{0}$, so that if P is, as usual, a positive polynomial with $\operatorname{Log} P = K \cap \mathbf{Z}^d$, then $1/P$ belongs to $R_K$, and $\phi$ induces an automorphism (as a ring) of $R_K[P] = \mathbf{R}[M(K),Q^{-1}]$, where the Q's vary over every positive polynomial with $\operatorname{cvx}\operatorname{Log} Q = mK$ for some integer m. Let v be a vertex of K which lies on a face containing the origin (which is also a vertex). Then the image of $x^v$ under $\phi$ must be an order unit times $x^v$. In other words, $\phi(x^v) = x^v(P_v/Q_v)$ where $P_v$, $Q_v$ are positive polynomials the convex hull of whose Log sets are nK for the same n. This extension of $\phi$ is determined by its values on these vertices, because every lattice point in W(K) is a positive rational linear combination of these particular v's.

We may have assumed from the beginning that K is d-dimensional, so that the field of fractions of $R_K$ is the function field $\mathbf{R}(x_1,\ldots,x_d)$. Then $\phi$ extends to an automorphism of this as well, and it is not hard to check that (particularly if K is integrally simple!) that its effect on the generators is of the following form:

$\phi(x_i) = x_i(f_i/g_i)$, where $f_i, g_i$ belong to $\mathbf{R}[x_i^{\pm 1}]^+$ and satisfy (*)

cvx Log $f_i$ = cvx Log $g_i$ = n(i)K.

Can we write down all such automorphisms? The conditions in (*) are very stringent, and moreover, since everything also applies to $\phi^{-1}$ the inverse transformation must have a similar form, say $\phi^{-1}(x_i) = x_i(h_i/j_i)$, with the n(i) changed to m(i). One source of such automorphisms is obtained from real characters of $(\mathbf{R}^d)^{++}$; specifically, let $r = (r_1,\ldots,r_d)$ be a strictly positive d-tuple of real numbers and define $\phi_r$ via:

$$\phi_r(x_i) = r_i x_i, \quad \text{and} \quad \phi^{-1}(x_i) = r_i^{-1} x_i.$$

It is easy to verify that with this definition we obtain a bona fide order-automorphism of $R_K$, and moreover the induced homeomorphism of $(\mathbf{R}^d)^{++}$ is just translation by r; on K, the induced homeomorphism is $\Lambda_P \Lambda_{\phi_r(P)}^{-1}$.

Are there any other automorphisms (that fix the states corresponding to the vertices)? For d=1, it is routine to check that there are no others, because all automorphisms of $\mathbf{R}(x)$ are fractional linear, and the only such satisfying (*) are given by $\phi_r$ for some r in (0,1). For d exceeding one, it is certainly plausible that such is the case, so we put it into the form of a conjecture about automorphisms of the field $D = \mathbf{R}(x_1, x_2,\ldots,x_d)$ (the reals could of course be replaced by any field):

> **CONJECTURE** Suppose that $\phi$ is an automorphism of D, the function field in d variables, such that $\phi$ (and its inverse) satisfy:
>
> $$\phi(x_i)/x_i = f_i/g_i,$$
>
> where the monomials appearing in $f_i$ are exactly the same as those appearing in $g_i$. Then $f_i/g_i$ is constant for all i.

To see that the hypotheses in the conjecture are implied by those of (*), just observe that if cvx Log $f_i$ = cvx Log $g_i$, there exists a polynomial Q such that Log $Qf_i$ = Log $Qg_i$, and if necessary this can be done for all i.

VIII.15 **PROPOSITION** Let K be an integrally simple d-dimensional indecomposable polytope. Then K is AGL(d,Z)-equivalent to n times the standard (unit) solid simplex in $\mathbf{R}^d$ for some integer n.

Proof: Let $\mathbb{K}$ denote the reduced form of K, and suppose (as we may without loss of generality) that $\mathbf{0} \in d_e K \subset (\mathbf{R}^d)^+$ and that the nearest lattice points to $\mathbf{0}$ in K along the edges thereof are $v_i = (0,\ldots,0,1,0,\ldots,0)$ with $i = 1,2,\ldots,d$. Set $P = \sum_{K \cap Z^d} x^w$. By VIII.10, 1/P is an irreducible element of $R_K$, and thus the ideal it generates is prime (since $R_K$ is factorial). It is an order ideal by II.2A, and it is a minimal prime ideal as it is principal. There thus exists a facet F of K such that

$$(1/P)R_K = \mathfrak{p}_F = \{x^w/P \mid w \in (\mathrm{Log}\ P)\backslash F\}.$$

Clearly $\mathbf{0}$ does not belong to F; we shall show that all of the $v_i$'s lie in F. If $v_i$ does not belong to F, then $a_j = x^{v_j}/P = x_j/P \in \mathfrak{p}_F = (1/P)R_K$. Hence $0 \leq a_j \leq N\cdot(1/P)$ for some positive integer N; in other words, the element $b = (N-x_j)/P$ belongs to $R_K$. However, this is impossible, since for $r_j > N$ and $r = (r_i)$ in $(\mathbf{R}^d)^{++}$, the value of b at r is negative.

Hence $v_j$ belongs to F. We conclude that F is the only facet of K that misses $\mathbf{0}$, and it follows that K is the standard simplex.∎

How good an AGL(d,Z)-invariant is $R_K$? On the basis of a limited number of examples, it is reasonably fine if K is indecomposable, but not so effective if K is not (see also Example VIII.19).

**EXAMPLE** Let $K_i$ ($i = 1, 2, 3, \ldots$) be the convex hull of $\{\mathbf{0}, (1,0), (0,i)\}$ inside $\mathbf{R}^2$; so each $K_i$ is a triangle, and $K_1$ is the standard 2-simplex; note that $K_2$ is the integral polytope arising from Example 1C. For simplicity of notation, let $R_i$ denote $R_{K_i}$. We shall show that the $R_i$ are mutually non-isomorphic, by examining the behaviour of the maximal ideals. We first note that $K_1$ is integrally simple, and none of the others are; so $R_1$ is factorial, but no $R_i$ is, for $i \geq 2$. We shall show that if $i \geq 2$, then each $R_i$ contains a unique maximal ideal $M_i$ such that $M_i/(M_i)^2$ is a real vector space of dimension i+1; for all other maximal ideals M, $M/M^2$ is 2-dimensional.

Fix i; let M be a maximal ideal of $R_i$, and define $P = \sum_{K \cap \mathbf{Z}^d} x^w$. Define the following elements of $R_i$: $a_1 = 1/P$, $a_2 = x/P$, and $a_3 = y^i/P$, corresponding to the three vertices of $K_i$. By the argument in either III.1 or III.2, at least one of $\{a_1, a_2, a_3\}$ does not belong to M. We first will show that if $a_1$ or $a_3$ belongs to M, then the localization at M, $(R_i)_M$ is a localization of a pure polynomial algebra, so that $M/M^2$ is 2-dimensional.

To this end, form $R_P = \mathbf{R}[a_1, a_2, a_3, y/P, y^2/P, \ldots, y^{i-1}/P]$, and recall that $R_i$ is obtainable from $R_P$ by inverting a multiplicative set. If $a_1$ does not belong to M, then $R_i[a_1^{-1}]$ is similarly obtainable from $R_P[a_1^{-1}]$. Now, $R_P[a_1^{-1}] = \mathbf{R}[1/P, P, x/P, y/P, y^2/P, \ldots, y^i/P] = \mathbf{R}[x, y, P^{-1}]$, a localization of $\mathbf{R}[x, y]$, so that $(R_i)_M$ is a localization of a pure polynomial algebra.

Next, consider $R_P[a_3^{-1}] = \mathbf{R}[P/y^i, y/P, y^2/P, \ldots, y^i/P, 1/P, x/P] = \mathbf{R}[y^{-1}, xy^{-i}, (P/y^i)^{-1}]$ (to see this, note that $P/y^i = y^{-1} + y^{-(i-1)} + \ldots + 1 + xy^{-i} \in \mathbf{R}[y^{-1}, xy^{-i}]$ ), so that $R_P[a_3^{-1}]$ is a localization of the pure polynomial algebra $\mathbf{R}[y^{-1}, xy^{-i}]$.

Hence for every maximal ideal M not containing either $a_1$ or $a_3$, $M/M^2$ is 2-dimensional. If both $a_1$ and $a_3$ belong to M, then for $0 < j < i$,

$$\left(\frac{y^j}{P}\right)^i = \left(\frac{y^i}{P}\right)^j \cdot \left(\frac{1}{P}\right)^{i-j} = a_3^j \cdot a_1^{i-j},$$

so that each of $y^j/P$ must also belong to M.

We have seen earlier (II.6) that if M is a maximal ideal of an $R_K$, then there exists a unique pure state $\gamma: R_K \longrightarrow \mathbf{R}$ such that $\ker \gamma = M$. For our choice of M now known to contain all of $\{y^j/P \mid 0 \le j \le i\}$, we see from $\sum_{K \cap Z^d} x^w/P = 1$, that $\gamma(y^j/P) = 0$, and $\gamma(x/P) = 1$. This uniquely determines the pure state $\gamma$ as $\gamma_{(1,0)}$, that is, the pure state corresponding to the vertex (1,0), and so uniquely determines the maximal ideal M. In particular, there is at most one maximal ideal M for which $M/M^2$ is of dimension other than 2. We proceed to show that for this M, the dimension of $M/M^2$ is exactly i+1.

Consider $R_P[a_2^{-1}] = \mathbf{R}[1/P, x/P, y^j/P, P/x] = \mathbf{R}[x^{-1}, yx^{-1}, y^2x^{-1}, \ldots, y^ix^{-1}, (P/x)^{-1}]$; thus $R_P[a_2^{-1}]$ is a localization of the monomial algebra $T = \mathbf{R}[x^{-1}; y^jx^{-1} \mid j = 1, 2, \ldots, i\}$. We observe that all of $x^{-1}$ and $y^jx^{-1}$ belong to $M(R_i)_M$, so that the corresponding quotient map (with kernel generated by M), $T \longrightarrow \mathbf{R}$ simply sends $1 \mapsto 1$, and all of $x^{-1}, y^jx^{-1} \mapsto 0$. It is easy to verify that $\{y^jx^{-1} \mid j = 0, 1, \ldots, i\}$ generate a maximal ideal, N, in T, and that $N/N^2$ has (real) basis $\{y^jx^{-1} + N^2 \mid j = 0, 1, \ldots, i\}$. It follows routinely that $M/M^2$ is i+1-dimensional.

In particular, we deduce that $\{R_i\}$ are mutually non-isomorphic. ∎

Now we consider the situation concerning order-automorphisms of $R_P$, at least where P is irreducible, as an element of $\mathbf{R}[x_i^{\pm 1}]$ (note that this plays the same rôle for polynomials that indecomposability does for polytopes).

Let us call an order-automorphism $\phi$ of $R_P$, <u>geometric</u> if there exists h in AGL(d,**Z**) such that for all w in Log P, $\phi(x^w/P)$ is a (real) scalar multiple of $x^{h(w)}/P$. Note that we do not assume that h(w) belong to Log P (as indeed it need not; it is only required that $x^{h(w)}/P$ be an element of $R_P$), although this will usually be the case. If $\phi$ is geometric, we shall see that there is a *unique* such element h, and there are additional constraints that apply to h, that are related to the coefficients of P.

VIII.16 **THEOREM** Let $P = \sum \lambda_w x^w$ be an irreducible element of $\mathbf{R}[x_i^{\pm 1}]$, having no negative coefficients. Then all order-automorphisms of $R_P$ (viewing the latter as an *ordered* ring) are geometric. In particular, $R_P$ has only finitely many order-automorphisms, and they are faithfully represented as a permutation group on the vertices of cvx Log P.

Some elementary preliminary results are required:

VIII.17 **LEMMA** If $P = \sum \lambda_w x^w$ is an irreducible element of $\mathbf{R}[x_i^{\pm 1}]$ having only positive coefficients, then the units of $R_P$ are trivial (that is, the only invertible elements of $R_P$ are scalars).

Proof: Let u be a unit; then for all pure states $\gamma$ of $R_P$, $\gamma(u) \neq 0$. As the pure state space is connected, one of $\pm u$ must be strictly positive as a function thereon, hence one of $\pm u$ is an order unit. We may obviously assume that u is itself an order unit of $R_P$. Thus we can write $u = g/P^m$ with g in $\mathbf{R}[x_i^{\pm 1}]^+$ and cvx Log P = m cvx Log P. Since $u^{-1} = P^m/g$ belongs to $R_P$, we can write $P^m/g = f/P^n$ for some f in $\mathbf{R}[x_i^{\pm 1}]$ and some integer n. This yields $fg = P^{m+n}$; as unique factorization holds in $\mathbf{R}[x_i^{\pm 1}]$, we must have that $g = \lambda P^s x^v$ for some positive real number $\lambda$, some integer s, and some v in $\mathbf{Z}^d$ (the units of $\mathbf{R}[x_i^{\pm 1}]$ are all of the form $\lambda x^v$ for some real number $\lambda$; positivity of g ensures that of $\lambda$). Thus Log g = w + s Log P. Hence

$$\text{cvx Log } g \;=\; w + s\,\text{cvx Log } P \;=\; m\,\text{cvx Log } P.$$

By equating the volumes (we may have assumed at the outset that P is projectively faithful, so that cvx Log P has nonzero d-volume), we deduce immediately that m = s. This now entails that $w = \mathbf{0}$ ($w + K = K$ implies $kw + K = K$ for all positive integers k; compactness yields $w = \mathbf{0}$). This simply says that $g = \lambda P^m$, so that u is the scalar $\lambda$.∎

While not directly relevant to the rest of the material in this section, the following is of interest:

VIII.17A **PROPOSITION** Let P be an element of $\mathbf{R}[x_i^{\pm 1}]^+$, and let $\{P_j\}_J$ be a complete set of irreducible factors of P. If there exist no relation of the form:

$$\sum_I n(i)\,\text{cvx}(\text{Log } P_i) \;=\; \sum_{I'} m(j)\,\text{cvx}(\text{Log } P_j)$$

where I and I' are disjoint subsets of J, and n(i) and m(j) are all positive integers, then $R_P$ has only trivial units.

Proof: Let $Q/P^m$ be a unit of $R_P$. Up to sign, it is an order unit, so we may assume that $Q$ lies in $R[x_i^{\pm 1}]^+$ and moreover $\text{cvx Log } Q = m\,\text{cvx Log } P$. Since $P^m/Q$ belongs to $R_P$, there exists a polynomial $Q'$ such that $QQ' = P^n$ for some integer $n$; thus all irreducible factors of $Q$ must appear among $\{P_j\}$. After clearing comon factors, we may write:

$$\frac{P^m}{Q} = \frac{\prod_I P_i^{n(i)}}{\prod_{I'} P_j^{m(j)}}$$

where $I$ and $I'$ are disjoint subsets of $J$. If $H$ is $P^m$ divided by the numerator (right side), then from $\text{cvx Log } Q = m\,\text{cvx Log } P$, we deduce:

$$\text{cvx Log } H + \sum_I n(i)\text{cvx Log } P_i = \text{cvx Log } H + \sum_{I'} m(j)\text{cvx Log } P_j.$$

From the cancellation lemma and the hypothesis, we deduce a contradiction.∎

VIII.18 **LEMMA** Under the hypotheses of VIII.17, for every $w$ in $\text{Log } P$, $x^w/P$ is an irreducible element of $R_P$.

Proof: Suppose it factors,

$$\frac{x^w}{P} = \frac{f}{P^k}\cdot\frac{g}{P^m}$$

for $f$ and $g$ elements of $\mathbf{R}[x_i^{\pm 1}]$, with $f/P^k$ and $g/P^m$ in $R_P$. We deduce that $fg = x^w P^{k+m-1}$, and irreducibility of $P$ yields the existence of scalars $\lambda$ and $\mu$, integers $s$ and $t$, and lattice points $v$ and $v'$, so that $f = \lambda x^v P^s$ and $g = \mu x^{v'} P^t$. Necessarily, $s + t = k + m - 1$, $v + v' = w$, and $\lambda\mu = 1$. As $f/P^k$ belongs to $R_P$, $\text{Log } f \subset k\,\text{cvx Log } P$, and thus $v + s\,\text{cvx Log } P \subset k\,\text{cvx Log } P$.

Taking volumes, we deduce that $s \leq k$; similar considerations lead to $t \leq m$. As $s + t = k+m-1$, we may assume that $s = k$ and $t = m-1$. This forces $f/P^k$ to be $\lambda x^v$; as $x^v$ is not bounded on $(\mathbf{R}^d)^{++}$ unless $v = \mathbf{0}$, we deduce that $f/P^k$ is the scalar $\lambda$. This means that the factorization is trivial, and so $x^w/P$ is irreducible.∎

Proof of VIII.16: Let $\phi$ be an order-automorphism of $R_P$. For $w$ in $\text{Log } P$, consider $z$ defined as $\phi(x^w/P)$. Since $(x^w/P)R_P$ is an order ideal, so is $zR_P$. There exists (since $z$ is positive) an element of the form $x^v/P^k$ (for $k$ a positive integer and with $v$ in $\text{Log } P^k$) such that

$0 \leq x^v/P^k \leq N\cdot z$, for some positive integer N. As $zR_P$ is an order ideal, there thus exists an element s of $R_P$ such that $x^v/P^k = z\cdot s$. Write, as usual,

$$z = f/P^m \quad \text{with} \quad \operatorname{Log} f \subset m \operatorname{Log} P;$$

$$s = f/P^n \quad \text{with} \quad \operatorname{Log} g \subset n \operatorname{Log} P.$$

We plug these into $x^v/P^k = z\cdot s$, and deduce as in the proofs of VIII.17 and VIII.18, that $f = \lambda x^a P^t$ (with a in $\mathbf{Z}^d$). As z is positive, $\lambda$ is a positive real number, and as $\operatorname{Log} f \subset m\operatorname{Log} P$, it follows that $a + t\operatorname{Log} P \subset m\operatorname{Log} P$. By taking the volumes of the convex hulls, we deduce $t \leq m$. If $t = m$, then $a = \mathbf{0}$, so that z would be a scalar. Thus $t < m$. We now wish to show that $t = m-1$.

We note that a belongs to cvx $(m-t)\operatorname{Log} P$ (by the cancellation lemma, just before VIII.10). Set $N = m-t$; we may express a as a *rational* convex combination of $\{Nv \mid v \in d_e \operatorname{cvx} \operatorname{Log} P\}$, say $a = \sum \mu_v N v$ with $\mu_v$ being non-negative rational numbers adding to 1. There exists an integer M so that for all v in $d_e$cvx Log P, $M\mu_v$ is an integer. From $Ma = \sum (NM\mu_v)v$, we deduce

$$z^M = \prod \left(\frac{x^v}{P}\right)^{NM\mu_v} .$$

Applying $\phi^{-1}$, we deduce

$$\left(\frac{x^w}{P}\right)^M = \prod \left(\phi^{-1}\left(\frac{x^v}{P}\right)\right)^{NM\mu_v} \qquad (1).$$

In the next paragraph, we shall show:

(††)

For every (ring) order-automorphism $\zeta$ of $R_P$, and every v in $d_e$Log P,

$$\zeta(x^v/P) = \lambda_v(x^{v'}/P)$$

for some v' in $d_e$Log P, and positive real $\lambda_v$.

Assume this for now; then applying it to (1), we deduce

$$\left(\frac{x^w}{P}\right)^M = \lambda_0 \prod_{v \in d_e \operatorname{Log} P} \left(\frac{x^{v'}}{P}\right)^{NM\mu_v} = \lambda_0 x^{\sum NM\mu_v v'} \cdot P^{-NM}.$$

Clearing denominators, we obtain:

$$x^{w'} = \lambda_o P^{M(1-N)} \quad \text{where} \quad w' = Mw - \sum MN\mu_v v'.$$

Thus $N = 1$, $\lambda_o = 1$, and $Mw = \sum MN\mu_v v'$, so $w = \sum N\mu_v v'$ (we are viewing $v'$ as a function of $v$). We deduce that $z = \lambda x^a/P$ (from $N = 1$), and necessarily (as $z$ is bounded), $a$ belongs to cvx Log P. Now we prove (††).

Substituting $\zeta = \phi$ and $v = w$ into the first paragraph of the proof, we deduce that $\zeta(x^v/P) = \lambda x^a/P^N$; the problem is that although $a$ belongs to cvx NLog P, it is not clear that $a$ belongs to NLog P. Now $\zeta$ induces an auto-homeomorphism of $d_eS(R_P,1)$, and as $\{\gamma_v \mid v \in d_e\text{Log } P\}$ are precisely the pure states whose kernels are order ideals, the action of $\zeta$ must be a permutation of the latter set; say $\gamma_{v'} \mapsto \gamma_v$. Applying $\gamma_{v'}$, we obtain $\gamma_{v'}(\zeta(x^v/P)) = \gamma_v(x^v/P) = 1/\lambda_v$ ($\lambda_v$ is the coefficient of $x^v$ in P). Hence $\gamma_{v'}(\lambda x^a/P^N) = 1/\lambda_v$ so $\gamma_{v'}(x^a/P^N) = \lambda/\lambda_v \neq 0$.

However, $\gamma_{v'}(x^a/P^N) = \lim_{t\to\infty} (t^{u\cdot a})/P^N(X(t))$, where $X(t)$ is the path in $(\mathbf{R}^d)^{++}$ given by $X(t) = (\dots, t^{u(i)}, \dots)$ and $u = (u(1), \dots, u(d))$ in $\mathbf{R}^d$ is a linear functional that exposes $v'$ as a vertex of cvx Log P. As $P(X(t)) = \sum \lambda_w t^{u\cdot w}$ and $u$ exposes $v'$, we deduce that $a = Nv'$ (as $a$ belongs to Ncvx Log P, otherwise the limit would be 0). Thus $\zeta(x^v/P)$ is an N-th power (of a scalar multiple) of $x^{v'}/P$; however, $x^v/P$ and thus its image under $\zeta$ are irreducible, which forces $N = 1$, so that $\zeta(x^v/P)$ is a scalar multiple of $x^{v'}/P$, proving (††).

The outcome of this is that for all $w$ in Log P, $\phi(x^w/P) = x^a/P$ (up to a positive scalar multiple) for some $a$ in $\mathbf{Z}^d \cap$ cvx Log P, and so that $x^a/P$ belongs to $R_P$. We can regard the $a$ as a function of the $w$. We now show that the assignment Log P $\longrightarrow \mathbf{Z}^d \cap$ cvx Log P given by $w \mapsto a$ extends to an element of AGL(d,Z).

For $w$ in $\mathbf{Z}^d$ such that $x^w/P^k$ belongs to $R_P$ (and so $w$ belongs to k cvx Log P), consider $\phi(x^w/P^k)$; we will obtain a function $a = a(w,k)$ with $a$ in $\mathbf{Z}^d \cap$ kcvx Log P, such that $\phi(x^w/P^k)$ is a scalar multiple of $x^a/P^k$ (notice that $P^k$ is not irreducible, so we cannot repeat the previous process exactly). There exists an integer M so that $Mw = \sum Mk\mu_v v$ (the sum taken over $v$ in $d_e$Log P) where the $\mu$'s are non-negative rational, adding up to 1, and $Mk\mu_v$ are all integers. Thus,

$$\left(\frac{x^w}{P^k}\right)^M = \prod_{v\in d_e\text{Log } P} \left(\frac{x^v}{P}\right)^{Mk\mu_v};$$

applying $\phi$,

$$z = \phi\left(\frac{x^{w}}{P^{k}}\right)^{M} = \lambda \prod \left(\frac{x^{v'}}{P}\right)^{Mk\mu_v} = \lambda . \frac{x^{\sum Mk\mu_v v'}}{P^{kM}} .$$

Writing $z = f/P^N$ with $f$ having no negative coefficients, and setting $v'' = \sum Mk\mu_v v'$, we have that $(f/P^N)^M = \lambda(x^{v''}/P^{kM})$; clearing denominators, $f^M = \lambda x^{v''}P^{(N-k)M}$. Thus (as $\lambda > 0$), $x^{v''}$ has an M-th root in $\mathbf{R}(x_i)$; as $\mathbf{R}[x_i^{\pm 1}]$ is integrally closed in $\mathbf{R}(x_i)$, $x^{v''}$ has an M-th root in the former, and it easily follows from the units being scalars time monomials, that $v''/M$ belongs to $\mathbf{Z}^d$. Thus $\sum k\mu_v v'$ belongs to $\mathbf{Z}^d$. We also deduce that $f = \lambda^{1/M} x^{v''/M} P^{N-k}$, so $z = \lambda^{1/M} x^{v''/M}/P^k$. Hence $a(w,k) = \sum k\mu_v v'$, which we can express as $\sum k\mu_v a(v,1)$. Up to this point, we have shown that if $x^w/P^k$ belongs to $R_P$ and $w = \sum_{d_e \text{Log} P} k\mu_v v$, (with $\mu_v$ being non-negative, rational, and their sum being 1), then

$$a(w,k) = \sum_{d_e \text{Log} P} k\mu_v a(v,1).$$

At this point, we may assume that P is projectively faithful, and that $\mathbf{0}$ belongs to Log P. Then $a(\mathbf{0},1)$ is such that $\phi(1/P) = \lambda(x^{a(\mathbf{0},1)}/P)$ for some positive real number $\lambda$. Let A be the semigroup $\cup_{n \in N} n\text{Log } P$; as $\mathbf{0}$ belongs to Log P, it follows that $A - A = \mathbf{Z}^d$. Define a candidate function $g: A \longrightarrow A$ via

$$g(w) = a(w,k) - k\,a(\mathbf{0},1) \qquad \text{whenever } x^w/P^k \text{ belongs to } R_P.$$

***Claim***: The candidate function g is a function, it is additive, and extends uniquely to an element of $GL(d,\mathbf{Z})$.

To this end, we first observe that if $x^w/P^k$ belongs to $R_P$, then for any integer $k'$ exceeding k, so does $x^w/P^{k'} = (x^w/P^k)(1/P)^{k'-k}$. Applying $\phi$ to this last equation, we deduce that $a(w,k') = a(w,k) + (k'-k)\,a(\mathbf{0},1)$. Thus $a(w,k) - k\,a(\mathbf{0},1)$ is independent of the choice of k (subject to our list of constraints up to this point, including $\mathbf{0}$ belonging to Log P, $x^w/P^k$ belonging to $R_P$, etc.), so that g is at least well-defined as a function on A.

Next, we must show that if w and w' are elements of A, then $g(w+w') = g(w)+g(w')$. Suppose $x^w/P^k$ and $x^{w'}/P^{k'}$ belong to $R_P$; then the product, $x^{w+w'}/P^{k+k'}$ also belongs, and we apply $\phi$; we deduce $a(w,k) + a(w',k') = a(w+w', k+k')$. Hence:

$$\begin{aligned} g(w+w') &= a(w+w', k+k') - (k+k')\,a(\mathbf{0},1) \\ &= a(w,k) - k\,a(\mathbf{0},1) + a(w',k') - k'\,a(\mathbf{0},1) \\ &= g(w) + g(w'). \end{aligned}$$

Thus g is a homomorphism. It is clear that the assignment associating such functions to order automorphisms of $R_P$, $\phi \mapsto g$, preserves composition, and so applying this to $\phi^{-1}$ yields an element g' such that $g \circ g'$ and $g' \circ g$ are both the identity. Hence $g: A \longrightarrow A$ is invertible, and since $\mathbf{Z}^d = A - A$, it easily follows that g extends uniquely to an automorphism of $\mathbf{Z}^d$, that is, to an element of $GL(d,\mathbf{Z})$, which we also call g. The corresponding element h in $AGL(d,\mathbf{Z})$ is given via $w \mapsto g(w) + a(\mathbf{0},1)$ for w in Log P. The definition of geometric is satisfied.

Next we show that the map $\phi \mapsto h \in AGL(d,\mathbf{Z})$ is one to one, that is if $\phi$ and $\phi'$ yield the same h, then $\phi = \phi'$ (note the difference between this situation and the case of automorphisms of $R_K$; the family of automorphisms parameterized by $(\mathbf{R}^d)^{++}$ given above would all induce trivial actions on cvx Log P). Consider the automorphism $\phi^{-1}\phi'$; this will have assigned to it the identity of $AGL(d,\mathbf{Z})$, so we need only show that if $\phi$ is an automorphism of $R_P$ such that $\phi(x^w/P)$ is a scalar multiple of $x^w/P$ for all w in Log P, then $\phi$ must be the identity automorphism.

Write, for each w in Log P, $\phi(x^w/P) = \rho_w \cdot (x^w/P)$, with $\rho_w$ being positive real numbers. From $\sum_{w \in \text{Log} P} \lambda_w (x^w/P) = 1$, we deduce after applying $\phi$, that $\sum_{w \in \text{Log} P} \lambda_w \rho_w (x^w/P) = 1$; multiplying by P,

$$\sum \lambda_w \rho_w x^w = \sum \lambda_w x^w.$$

As this is an identity of polynomials, we deduce immediately that $\lambda_w \rho_w = \lambda_w$ for all w in Log P, and thus $\phi(x^w/P) = x^w/P$ for all such w. As the set $\{x^w/P \mid w \in \text{Log } P\}$ generates $R_P$ as a real algebra, we deduce that $\phi$ must be the identity as desired.

Thus $\phi$ is uniquely determined by the effect of h on cvx Log P, and it follows from the equation (1) above that $\phi$ is uniquely determined by its effect on terms of the form $x^v/P$ with v running over the vertices of cvx Log P. Correspondingly, $\phi$ is uniquely determined by the effect of h on the vertices, and so the last portion of the statement of VIII.16 is proved.∎

Given h in $AGL(d,\mathbf{Z})$ and $P = \sum \lambda_w x^w$ in $\mathbf{R}[x_i^{\pm 1}]^+$, under what conditions will h induce an automorphism of P? It is not necessary that P be irreducible in what follows. If such an automorphism, $\phi$ exists, then for all w in Log P, it must be that $\phi(x^w/P) = \rho_w \cdot (x^{h(w)}/P)$ for some positive real number $\rho_w$, and moreover, if w is a vertex of K = cvx Log P, then so is h(w); also h(K) = K is necessary. As in the proof above, $1 = \sum \lambda_w (x^w/P)$ leads to $\sum \lambda_w \rho_w x^{h(w)} = P = \sum \lambda_w x^w$. We observe that this forces $h(\text{Log } P) \subset \text{Log } P$. If for some w' in Log P, h(w') does not belong to Log P, then using the equality of the coefficients we deduce that $\lambda_{w'} \rho_{w'} = 0$; however, neither $\lambda_{w'}$ nor $\rho_{w'}$ can be zero.

Thus $\lambda_{h(w)}$ makes sense, and from $\sum \lambda_w x^w = \sum \lambda_{h(w)} x^{h(w)}$, we deduce that for all w in Log P, $\lambda_w \rho_w = \lambda_{h(w)}$, or in other words, $\rho_w = \lambda_{h(w)}/\lambda_w$. There is another necessary condition, arising from multiplicativity of $\phi$ (e.g., if v + v' = w + w' for v, v', w, w' in

Log P, then $\rho_v \rho_{v'} = \rho_w \rho_{w'}$) which assert affectively that there exists a group homomorphism $\chi:(\mathbf{R}^d,+) \longrightarrow (\mathbf{R}^{++},\cdot)$ and a positive real number $\rho$ such that $\rho_w = \rho\cdot\chi(w)$. These conditions are quite stringent. On the other hand if h in AGL(d,Z) satisfies h(Log P) = Log P, and $\lambda_w = \lambda_{h(w)}$ for all w in Log P, then it does induce an automorphism of $R_P$ given by $x^w/P \mapsto x^{h(w)}/P$.

If $P = 1 + x + x^3 + x^4 + x^5$, then $x^2/P$ belongs to $R_P$ (consider $R_{P^2}$!), so that $R_P$ admits no order-automorphisms (notice that h(3) would be forced to equal 2). If $P = ax^3 + bx^2 + cx + 1$ (with a, b, c positive real numbers), then a necessary (and apparently sufficient) condition for there to a be non-trivial (ring) order-automorphism is that $ab^3 = c^3$; the corresponding automorphism must send $x^3/P$ to $(1/a)(1/P)$ and $x^2/P$ to $(c/b)(x/P)$.

VIII.19 **EXAMPLE** In contrast to the examples of triangles and their corresponding $R_K$'s described just after VIII.15, here we describe two quadrilaterals, K and K' that are not even AGL(d,**R**)-equivalent, yet such that $R_K$ is isomorphic to $R_{K'}$. Several other points will also be illustrated.

Let K be the unit square in $\mathbf{R}^2$ and let K' = cvx{**0**, (1,0), (2,0), (0,1), (1,1)}. We shall prove that $R_K$ is isomorphic to $R_{K'}$; actually we show that if P = (1+x)(1+y) and P' = (1+x+y)(1+x), then $R_P$ is isomorphic to $R_{P'}$. We also prove that the elements 1/(1+x+y) and x/(1+x+y) are *not* irreducible (either in $R_{P'}$ or $R_{K'}$); this is the obstruction to extending the arguments in the proofs of VIII.12 and VIII.16 to the decomposable (respectively, reducible) situations. In this particular case, K' is integrally simple, so that VIII.13 applies anyway.

Set Z = 1/(1+x+y), X = x/(1+x+y), W = y/(1+x+y), Y = 1/(1+x), and V = x/(1+x); all of these belong to $R_{P'}$. Using partial fractions, we see that $R_{P'} = \mathbf{R}[V,W,X,Y,Z]$; as Z + X + W = 1 and Y + V = 1, $R_{P'} = \mathbf{R}[X,W,V]$. However,

$$X = \frac{1+x}{1+x+y}\cdot\frac{x}{1+x} = (1-W)V.$$

Thus, $R_{P'} = \mathbf{R}[W,V]$; as as $R_{P'}$ has transcendence degree 2, there can be no relations among W and V. Thus $R_{P'}$ is at least a pure polynomial algebra. Now we consider the order structures on $R_P$ and $R_{P'}$.

Suppose that $P_1$ and $P_2$ belong to $\mathbf{R}[x_i^{\pm 1}]$ and $Q = P_1P_2$. By definition, $R_Q^+ = \langle\{x^w/Q \mid w \in \text{Log } Q\}\rangle$; we show in fact that

$$R_Q^+ = \langle\{x^w/P_1 \mid w \in \text{Log } P_1\} \cup \{x^w/P_2 \mid w \in \text{Log } P_2\}\rangle. \qquad (*)$$

Obviously, the right hand side is contained in the left. If $w$ belongs to Log Q, then $w = w_1 + w_2$ for $w_i$ in Log $P_i$ (as $P_1$ and $P_2$ have no negative coefficients), so that $x^w/Q = (x^{w_1}/P_1)\cdot(x^{w_2}/P_2)$, so the equality holds.

With $P_1 = 1 + x + y$ and $P_2 = 1 + x$, it follows that $R_{P'}^{+} = \langle V, W, X, Y, Z \rangle$. Note that $1-W = X+Z \in R_{P'}^{+}$ and $1-V = Y$; as $X = (1-W)V$ (as previous) and $Z = (1-V)W$ (easily verified), we see that

$$R_{P'}^{+} = \langle 1-W, W, 1-V, V \rangle.$$

Again from (*) with $P_1 = 1+x$ and $P_2 = 1+y$, we see that $R_P = \mathbf{R}[1/(1+x), 1/(1+y)]$ (the pure polynomial algebra in $(1+x)^{-1}$ and $(1+y)^{-1}$), and

$$R_P^{+} = \left\langle \frac{x}{1+x}, \frac{1}{1+x}, \frac{y}{1+y}, \frac{y}{1+y} \right\rangle.$$

The assignment $W \mapsto 1/(1+x)$, $V \mapsto 1/(1+y)$ extends uniquely to an isomorphism of rings, $\phi: R_{P'} \longrightarrow R_P$ and from the expressions for the positive cones, $\phi$ sends $R_{P'}^{+}$ onto $R_P^{+}$, and moreover, $\phi^{-1}$ maps $R_P^{+}$ onto $R_{P'}^{+}$; thus $\phi$ is an order-isomorphism.

Since $\phi$ restricts to a bijection between the sets of order units of $R_{P'}$ and of $R_P$ it thus extends to an (order)-isomorphism $R_{K'} \longrightarrow R_K$.

The equations $X = (1-W)V$ and $Z = (1-V)W$ yield that neither $x/(1+x+y)$ nor $1/(1+x+y)$ is irreducible in $R_{P'}$ or $R_{K'}$.

Finally, we note that the order automorphism of $R_{P'}$ given by $W \mapsto 1-W$, $V \mapsto 1-V$ is *not* geometric, while the corresponding automorphism of $R_P$ is.∎

## Appendix A. CALCULATION OF PIC ($R_P$) FOR CHONDROPHAGIC P

In this appendix, we show how to calculate the Picard groups of $R_P$'s for those positive polynomials for which $R_P$ is homologically regular (see section VI, especially VI.6). These are precisely the P's for which the ring obtained by inverting all of the order units of $R_P$ is factorial; equivalently, $\cup K_v \subset \text{Log } P \subset K = \text{cvx Log } P$, the v's running over the vertices of K (see VI.6). Inasmuch as all projectives are free over $R_K$, it turns out not to be very difficult to calculate Pic($R_P$), and this is intimately related to the factorization of P within $\mathbf{R}[x_i^{\pm 1}]$ and the geometric structure of cvx Log P. In particular, if $R_P$ is regular, then Pic($R_P$) is at most e–d-generated, where e is the number of facets of K. If P is additionally irreducible (as an element of $\mathbf{R}[x_i^{\pm 1}]$), then $\text{Pic}(R_P) = \mathbf{Z}^{e-(d+1)} \oplus \mathbf{Z}/a\mathbf{Z}$ for some integer $a \geq 1$.

We also show that generically, $R_P$ is not a unique factorization domain; specifically, if P in $\mathbf{R}[x_i^{\pm 1}]^+$ is irreducible as an element of $\mathbf{R}[x_i^{\pm 1}]$, and P is projectively faithful (to which form P can always be reduced in any case), then $R_P$ being factorial implies that up to the natural action of AGL(d,**Z**) on Log P, P has the extremely simple form, $P = \lambda_o + \sum \lambda_i x_i$ (the λ's being positive real numbers of course)!

We first require some results of a more general nature about $R_P$ and $S_P = \mathbf{R}[x_i^{\pm 1}, P^{-1}]$ (the latter with the limit ordering) leading up to finding generators for the Picard group.

**A1 LEMMA** Let P be an element of $\mathbf{R}[x_i^{\pm 1}]^+$, and define $S_P = \mathbf{R}[x_i^{\pm 1}, P^{-1}]$ with the limit ordering. For any w in $\mathbf{Z}^d$, for any k in **Z**, $x^w P^k R_P$ is an order ideal of $S_P$ (viewing the latter as a partially ordered group).

> **REMARK.** Order ideals in ordered rings are not generally ring ideals; however, if 1 is an order unit for the ordered ring, then order ideals are ideals (see the comment just after I.1).

<u>Proof:</u> Clearly $I = x^w P^k R_P \subset S_P$, and the relative ordering from $S_P$ agrees with that obtained from $R_P$, on $x^w P^k R_P$. If a belongs to I, $a = x^w P^k a_o$ for some $a_o$ in $R_P$, and as $R_P$ has an order unit, it follows that $x^w P^k$ is an order unit for I, and thus I is directed. If $0 \leq a \leq b$ with b in I and a in $S_P$, we see that

$$0 \leq x^{-w} P^{-k} a \leq x^{-w} P^{-k} b = b_o \in R_P$$

(as multiplication by a monomial, $x^w$, or by a power of P, is an order-automorphism of $S_P$, when we consider the latter only as a partially ordered group. Since $R_P$ is the order ideal of $S_P$ generated by 1, $x^{-w} P^{-k} a$ belongs to $R_P$, so a belongs to I, and thus I is convex.∎

**A.2 COROLLARY** If $J$ is an order ideal of $R_P$ that is finitely generated as an $R_P$-module by a set of elements each of which has the form $x^w/P^k$ ($w \in \mathbf{Z}^d$, $k \in \mathbf{Z}$), then $J{:}S_P = \{s \in S_P \mid sJ \subset R_P\}$ is an order ideal of $S_P$.

Proof: Write $J = \sum x^{w_i}P^{k_i}R_P$; then $J{:}S_P = \bigcap x^{-w_i}P^{-k_i}R_P$; by the lemma, each $x^wP^kR_P$ is an order ideal, and in a dimension group, finite intersections of order ideals are order ideals.∎

**A.3 LEMMA** If $I$ and $J$ are order ideals of $S_P$, and are finitely generated as $R_P$-modules by terms of the form $x^w/P^k$, then the product, $IJ$, is an order ideal of $S_P$, finitely generated in the same fashion.

Proof: Clearly $IJ$ is so generated as an $R_P$-module. It remains to show that any (finite) sum of principal $R_P$-modules (as in A.1) is itself an order ideal. In a dimension group, any sum of order ideals is an order ideal.∎

We now observe that if $J$ is an $R_P$-module contained inside $S_P$, or more generally in the field of fractions of $S_P$, then we may define $J^{-1}$ as $\{e = ab^{-1} \mid a,b \in R_P,\ eJ \subset R_P\}$. If $J = x^wP^kR_P$, then clearly $J^{-1} = x^{-w}P^{-k}R_P = J{:}S_P$; thus if $J$ is as described in A.2, then $J^{-1} = J{:}S_P$. A (fractional) ideal $J$ of a commutative ring $R$ is said to be invertible if $JJ^{-1} = R$; this is equivalent to $J$ being projective as an $R$-module.

**A.4 LEMMA** If $I$ and $J$ are order ideals of $R_P$ (hence of $S_P$) with $I \subset J$, then $IJ^{-1}$ is an order ideal of $R_P$.

Proof: By [H1; I.I(a)], $I$ and $J$ are ideals of $R_P$, so that $IJ^{-1}$ is an ideal, hence an $R_P$-submodule of $S_P$. By the previous results, $IJ^{-1}$ is an order ideal of $S_P$, and being contained in $R_P$, it is an order ideal of the latter.∎

**A.5 LEMMA** If $R$ is a commutative integrally closed domain, and $I$ is an invertible ideal of $R$, then $I$ has no embedded primes.

Proof: At every localization at a prime ideal, $I$ is principal, so the latter has no embedded primes, and it follows that $I$ has none either (just localize at a maximal ideal suspected of containing an embedded prime of $I$).∎

A.6 **COROLLARY** If R is a noetherian integrally closed domain such that every minimal prime ideal is projective, and I is a nonzero projective ideal then $I = \prod \mathfrak{p}_i^{k(i)}$, the $\mathfrak{p}_i$ running over all the minimal primes containing I, with k(i) being positive integers.

Proof: Let $\{\mathfrak{p}_i\}$ be the finite set of minimal prime ideals containing I. Define k(i) to be the least integer such that $I \subset \mathfrak{p}_1^{k(1)}$, and set $I_1 = I\mathfrak{p}_1^{-k(1)} \subset R$. Then $I_1$ is projective and satisfies the same hypotheses as I, except that the former is not contained in $\mathfrak{p}_1$ but is contained in all the rest of the $\mathfrak{p}_i$'s, but no other minimal prime ideals. By the obvious induction process, we reduce to the case that I is contained in only one minimal prime, $\mathfrak{p}$. Then $J = I\mathfrak{p}^{-1}$ is a projective ideal contained in no minimal prime ideals; as J has no embedded primes, J = R, and we deduce $I = \mathfrak{p}^k$, allowing the induction to commence.∎

For a ring (here a commutative domain with 1), R, Pic(R) [Ba; p.71], [Sw2; §8] is defined as the group of isomorphism classes of invertible ideals with operation defined by [I] + [J] = [IJ] (it is more convenient in what follows to use additive notation, rather than the more usual multiplicative). There is a group homomorphism [Sw2; p. 146] $\det: K_0(R) \longrightarrow \mathrm{Pic}(R)$ such that for all invertible ideals I of R,

$$\det\left([I]_{K_0(R)}\right) = \det\left([I]_{K_0(R)} - [R]_{K_0(R)}\right) = [I]_{\mathrm{Pic}(R)}.$$

In other words, if I and J are projective ideals that are stably isomorphic, then they are isomorphic. Hence if Pic(R) = 0 and R is regular, then R is factorial. On the other hand, if R is factorial, then Pic(R) = 0.

Following Bass, let T be a multiplicatively closed set inside the (homologically) regular ring R. Then T is "factorial for R" [Ba; p. 144] and if $j_T: R \longrightarrow T^{-1}R$ is the natural map, then

(i) [Ba; p. 144, 7.17] $\mathrm{Pic}(j_T): \mathrm{Pic}(R) \longrightarrow \mathrm{Pic}(T^{-1}R)$ is surjective;

(ii)[Ba; p. 136, 7.10] the kernel of $\mathrm{Pic}(j_T)$ is generated by

$$\{[\mathfrak{p}] \mid \text{minimal prime ideals with } \mathfrak{p} \cap T \text{ being non-empty}\}.$$

Now let P in $\mathbf{R}[x_i^{\pm 1}]^+$ satisfy the hypotheses of VI.6; let K denote cvx Log P. We shall call such a P, a chondrophage[1]. Then $R = R_P$ is regular by VI.6. There exists an integer n so that $n\mathrm{Log}P \cap \mathrm{Int}(nK)$ is non-empty; let w be an element therein. By II.1, $R_P[(x^w/P^n)^{-1}] = S_P =$

[1]This word is as close as I could get to "fibre-eater" using a word of Greek origin; I wanted a synonym for the over-used "regular".

$R[x_i^{\pm 1}, P^{-1}]$, so if $T = \{(x^w/P^n)^m | m = 0,1,2,...\}$, then T is a multiplicative set and $\mathrm{Pic}(T^{-1}R_P) = \mathrm{Pic}(S_P) = 0$. So $\mathrm{Pic}(R_P)$ is generated by $\{[\mathfrak{p}] | \mathfrak{p}$ is a minimal prime ideal containing $x^w/P^n\}$. Since $(x^w/P^n)R_P$ is an order ideal, by II.2A, all minimal primes containing it are themselves order ideals, and thus must be of the form, $\mathfrak{p}_F$, generated as an ideal by $\{x^v/P | v \in \mathrm{Log}\, P \setminus F\}$ for a *facet* F (the latter in order to be minimal as a prime ideal). We thereby deduce that $\{[\mathfrak{p}_F] | F$ a facet of $K\}$ generates $\mathrm{Pic}(R_P)$. Now we find all the relations among them.

Let e be the number of facets of K, and index the facets $F_1, ..., F_e$, with corresponding minimal prime ideals $\mathfrak{p}_1, ..., \mathfrak{p}_e$. Let v be an element of $\mathrm{Log}\, P \cap \mathbf{Z}^d$, and consider the ideal $(x^v/P)R_P$; this is an order ideal, so the minimal primes containing it are all of the form $\mathfrak{p}_i$ (V.4), and as the latter are projective and $R_P$ is integrally closed, by A.6, $(x^v/P)R_P = \prod_{1\le i\le e} \mathfrak{p}_i^{k(i)}$. This yields a lot of relations among the generators of the Picard group.

Since K is integrally simple, there exists an element of $\mathrm{AGL}(d,\mathbf{Z})$ whose application to K allows us to assume that $K \subset (\mathbf{R}^d)^+$ and $\mathbf{0}$ belongs to $d_e K$, and furthermore, that the peak polytope $K_v$ with $v = \mathbf{0}$ (we obviously cannot write $K_0$!) is the standard unit simplex. Relabel the d facets containing $v = \mathbf{0}$ as $F_1, ..., F_d$. For each w in $K_v \cap \mathbf{Z}^d$, we obtain a relation among the $[\mathfrak{p}_i]$. The factorization $(x^w/P)R_P = \prod \mathfrak{p}_i^{k_w(i)}$ yields:

$$0 = \sum_{i=1}^{e} k_w(i)[\mathfrak{p}_i] \qquad (k_w(i) \in \mathbf{Z}^+).$$

There is an elementary geometric way of deciding what the $k_w(i)$ are, given w. Let $\alpha_i$ be the affine linear functional describing the facet $F_o$, that is, $\alpha_i | F_i = r_i$ and $\alpha_i | K \ge r_i$. Let $a_i$ be the smallest value of $\alpha_i$ on $(K \setminus F) \cap \mathbf{Z}^d$, and set $b_i = a_i - r_i$. Then $k_w(i) = (1/b_i)(\alpha_i(w) - r_i)$. To see this, we may assume that $F = \{\mathbf{0} \times \mathbf{R}^{d-1}\} \cap K$, $K \subset (\mathbf{R}^d)^+$, and that $\bigcup nK = (\mathbf{R}^d)^+$ (since K is integrally simple); then $\alpha$ is just the projection onto the first coordinate, the corresponding r is zero, and $a = b = 1$. It is clear that $\alpha(w) = s$ if and only if $x^w/P$ belongs to $\mathfrak{p}_F^s \setminus \mathfrak{p}_F^{s+1}$.

With our choice of $v = \mathbf{0}$, $k_v(i) = 0$ for $i = 1, 2, ..., d$, but $k_v(i) > 0$ for $i > d$. For $w = (0, ..., 0, 1, 0, ..., 0)$ (the 1 in the j-th position), $k_w(i) = 1$ for exactly one of the $\mathfrak{p}_i$'s with $i \le d$, call it $\mathfrak{p}_j$, and zero for the other $\mathfrak{p}_i$'s with $i \le d$. If $i > d$, then $k_w(i)$ is nonzero for at most one i, and then only if w also happens to be a vertex of K.

Let B denote the $e \times (d+1)$ matrix given by $B_{ij} = k_w(i)$ where $w = (0, ..., 0, 1, 0, ..., 0)$ (the 1 in the position $j-1$) if $j > 1$, and $w = \mathbf{0}$ if $i = 1$. Then there is a natural onto map:

$$\mathrm{Cok}(B) = \mathbf{Z}^e/B(\mathbf{Z}^{d+1}) \longrightarrow \mathrm{Pic}(R_P)$$

which is obtained from the asignment, for $f_j = (0, ..., 0, 1, 0, ..., 0)$ belonging to $\mathbf{Z}^e$,

$$f_j + \mathrm{Im}(B) \mapsto [\mathbf{p}_j].$$

Now the matrix B has the following form:

$$B = \begin{bmatrix} 0 & 0 & \dots & 0 & * \\ 1 & 0 & \dots & 0 & * \\ 0 & 1 & \dots & 0 & * \\ & & \dots & & \\ 0 & 0 & \dots & 1 & * \\ * & * & \dots & * & * \end{bmatrix}.$$

Thus the rank of B is either d or d+1; since the first row is not zero (see next sentence), the rank is exactly d+1. Moreover, B is row equivalent (via GL(e,Z)) to

$$B' = \begin{bmatrix} 0 & 0 & \dots & 0 & a \\ 1 & 0 & \dots & 0 & * \\ 0 & 1 & \dots & 0 & * \\ & & \dots & & \\ 0 & 0 & \dots & 1 & * \\ 0 & 0 & \dots & 0 & 0 \end{bmatrix},$$

where a is the gcd of the entries of the last column of B, and is thus the greatest common divisor of the nonzero members of $\{k_v(i) \mid i \le e\}$ (with $v = \mathbf{0}$). Thus $\mathrm{Cok}(B) = \mathbf{Z}^{e-(d+1)} \oplus (\mathbf{Z}/a\mathbf{Z})$. We thereby deduce

1. $\mathrm{Pic}(R_P)$ is e–d generated, and its torsion-free rank is at most e–d–1.

Now consider a relation on the generators $[\mathbf{p}_i]$ obtained from an *arbitrary* w in $n \operatorname{Log} P$, i.e., if

$$(x^v/P)R_P = \prod \mathbf{p}_i^{k_{w,n}(i)}$$

(the preceding computations only involved n=1), we obtain the relation in $\mathrm{Pic}(R_P)$:

$$0 = \sum k_{w,n}(i)\,[\mathbf{p}_i].$$

We claim this relation is a consequence of the d+1 relations given previously by the matrix B. Assuming as always that $v = \mathbf{0}$ and $K_v$ is the standard unit simplex, we may write w =

$\Sigma_{1\le i\le d}$ w(i) $v_i$ (where the $v_i$ are the standard basis elements of $Z^d$, and these are the nonzero vertices of $K_v$). Then

$$\left(\frac{1}{P}\right)^{\sum w(i)-n} \cdot \left(\frac{x^w}{P^n}\right) R_P = \left(\prod \left(\frac{x^{v_i}}{P}\right)^{w(i)}\right) R_P \qquad \text{if } \sum w(i) \ge n.$$

and

$$\frac{x^w}{P^n} = \prod \left(\frac{x^{v_i}}{P}\right)^{w(i)} \cdot \left(\frac{1}{P}\right)^{n-\sum w(i)} \qquad \text{if } \sum w(i) < n.$$

In either case (using fractional ideals if necessary), we see that the relation obtained from w is a consequence of those obtained from the first d+1 relations.

Now we examine other sources of relations among the generators of Pic($R_P$). If $I = \prod \boldsymbol{p}_i^{k(i)}$ is a principal ideal, it being an order ideal as well, we may write $I = zR_P$ with z in $R_P^+$ which is an order unit for I (V.1). Suppose that $x^w/P^m$ belongs to I (as I is an order ideal, it is generated by such elements); then we may write $z = f/P^n$ with $\text{Log } f \subset n\text{Log } P$ and f in $\mathbf{R}[x_i^{\pm 1}]^+$, and we may find an element $g/P^k$ with $\text{Log } g \subset k \text{ Log } P$ such that $x^w/P^m = (f/P^n)(g/P^k)$. In other words $f g x^{-w} = P^{n+k-m}$ (so $n+k \ge m$). If P is irreducible as an element of $\mathbf{R}[x_i^{\pm 1}]$, then we deduce that up to multiplication by monomials in x, f and g are powers of P, and we derive a relation among the equivalence classes $[\boldsymbol{p}_i]$ of the form considered earlier. Thus we have:

2. If P is irreducible as an element of $\mathbf{R}[x_i^{\pm 1}]$, (and is a chondrophage), then

$$\text{Pic}(R_P) = \mathbf{Z}^{e-(d+1)} \oplus \mathbf{Z}/a\mathbf{Z} = \text{Cok}(B).$$

Nowwe can show that all additional relations arise from those generated by the factorization of P into irreducibles in $\mathbf{R}[x_i^{\pm 1}]$. In other words, if $f g x^{-w} = P^b$, then the relation arising from the factorization obtained above (with $b = n+k-m$) is a consequence of the finite set of relations obtained from factorizations of P itself, together with those arising from the matrix B.

Let $P_j$ be one of the (irreducible) factors of P. From it, we obtain the factorization (as $(P_j/P)R_P$ is an order ideal),

$$\left(\frac{P_j}{P}\right) R_P = \prod_i \boldsymbol{p}_i^{k(i,j)}.$$

If $Q = \prod P_j^{t(j)}$, and (say) $\text{Log } Q = m\text{Log } P$, then:

$$\prod_j \left(\frac{P_j}{P}\right)^{t(j)} = \left(\frac{Q}{P^m}\right)\cdot\left(\frac{1}{P}\right)^{\sum t(j)-m};$$

(If $m > \sum t(j)$, put the power of $1/P$ on the left.) It follows from unique factorization for projective ideals that the relation generated by Q is up to a principal ideal and a monomial given by:

$$\sum_j \sum_i t(j)\,k(i,j)\,[\mathbf{p}_i] = 0,$$

and this is a **Z**-linear combination of the relations arising from the irreducible factors.

If for example, $\mathrm{cvx\,Log}\,P_1 = \mathrm{cvx\,Log}\,P_2$, then

$$\frac{P_1}{P_2} = \frac{P_1 P/P_2}{P} \in R_P$$

(recall that $R_P$ is integrally closed), and it is easy to see that $P_2/P_1$ is an order unit, since $P_2/P_1$ also belongs to $R_P$. It follows that $P_1$ and $P_2$ yield the same relations among the $[\mathbf{p}_i]$.

More generally, if Q and Q' divide powers of P, and cvx Log Q and cvx Log Q' are homothetic to each other, then up to the relations generated by the monomials, the two relations they yield are (rationally) commensurable. Thus to determine at least the torsion-free rank of $\mathrm{Pic}(R_P)$, we start with a list of irreducible factors; as the list is built, discard any irreducibles which (possibly in combination with others already on the list) yields a homothety of the kind described above. Then the torsion-free rank is e–d–(# of elements in the list). An amusing application for this is the following, which presumably has been proved in the more general context of simple (real) polytopes. For this reason we offer a detailed sketch of the proof rather than a complete proof:

**A.7 THEOREM** Let K be a d-dimensional integrally simple polytope with e facets. Then there exists an integer n so that nK can be expressed as

$$\sum_{i=1}^{e-d} a_i K_i$$

where the $a_i$ are strictly positive integers and $\{K_i\}_{i=1}^{e-d}$ is a collection of $e-d$ pairwise non-homothetic indecomposable integral polytopes.

<u>Proof</u>: Of course $\mathrm{Pic}(R_K) = 0$, so given an order of the form $R_P$ in $R_K$, there exists P' in $\mathbf{R}[x_i^{\pm 1}]^+$ such that $R_P \subset R_{PP'} \subset R_K$ and $\mathrm{Pic}(R_{PP'}) = 0$, and any intermediate ring of the form $R_Q$ (containing $R_{PP'}$) also has trivial Picard group (since the Picard group is generated by the ideal classes of the minimal prime order ideals; these correspond to the facets).

If $K$ is already indecomposable, then it is a simplex by VIII.15, so there is nothing to do. If not, $K$ decomposes into a non-trivial sum of integral polytopes, say $K = K_1 + K_2$; unfortunately, these summands need not be integrally simple (see Example A-2), so we cannot work inductively to obtain $K$ itself as a sum of the correct number of indecomposables. However, some integer multiple of $K_1$ will admit a proper indecomposable summand (or else $K_1$ itself would be indecomposable). Since rational polytopes decompose into rational indecomposable polytopes, given an indecomposable $K'$ appearing in a summand of a multiple of $K$, there exist integral polytopes $K_i'$ which are indecomposable, and an integer $a$ so that $aK'$ is integral, such that $aK' + \sum a_i' K_i' = mK$ for some integers $a_i$, and $m$.

Let $\{K_i\}_{i=1}^{i=t}$ be the collection of all indecomposable, reduced (up to homothety) polytopes that appear as summand of multiples of $K$; if $t < e-d$ (in fact, not only is $t < \infty$, but $t \leq e-d$ in general by [McM; Cor. p. 94]), we shall obtain a contradiction.

Let $Q$ in $\mathbf{R}[x_i^{\pm 1}]^+$ be such that $R_K$ is attainable by inverting all of the order units of $R_Q$; by enlarging $Q$ if necessary, we may assume $\text{cvx Log } Q = nK$. Factor $Q = \prod Q_j$. Replacing $Q$ by a power of itself, each $\text{cvx Log } Q_j$ may be assumed to be a sum of indecomposable reduced polytopes, that is, certain of the $K_i$ (up to translation). For each $K_i$, let $P_i = \sum x^w$, the sum taken over $K_i \cap \mathbf{Z}^d$. Set $P = \prod P_i^{f(i)}$, with the $f(i)$ chosen so that $\sum_{1 \leq i \leq t} f(i) K_i = mK$. Form $R_{PQ}$; of course, this is still an order in $R_K$, and if we had chosen $Q$ so that $R_Q$ has trivial Picard group (as we may) then $\text{Pic}(R_{PQ}) = 0$ as well, by the earlier comment relating the facets to the generators of the Picard group.

For every irreducible factor of $PQ$, the convex hull of its Log set (up to translation) is a sum of multiples of certain of the $K_i$'s. Hence the relation among the ideal classes that is deduced therefrom is a consequence of the relations arising from the $K_i$'s. Since $\text{Pic}(R_{PQ})$ is $\text{Cok}(B)$ modulo these relations and the torsion-free rank of $\text{Cok}(B)$ is $e-d-1$, we must have a set of $e-d-1$ linearly independent relations. Now the $t$ indecomposables give rise to a rational vector space of dimension at most $t-1$: From $mK = \sum f(i) K_i$, the sum of the relations is equal to that obtained from $1/P$; this has already been incorporated into $\text{Cok}(B)$. Hence $t-1 \geq e-d-1$, so $t \geq e-d$, as desired.∎

Note that the proof also yields that if among the factors of the chondrophage $P$ are $P_i$ such that $\text{cvx Log } P_i$ exhaust the indecomposable summands of $K$, then $\text{Pic}(R_P)$ is finite.

**EXAMPLE** A-1. Let K be the standard d-simplex, and suppose P is an element of $\mathbf{R}[x_i^{\pm 1}]^+$ such that cvx Log P = nK, and $\cup\, (nK)_v \subset$ Log P for all v in $d_eK$ (viz. VI.6). We calculate $\mathrm{Pic}(R_P)$.

Let $P = \prod P_i$ be the factorization into (not necessarily distinct) irreducibles. It is easy to see that cvx Log $P_i$ = n(i)K for some integers n(i) (with $\sum n(i) = n$). We claim that if b = gcd {n(i)}, then $\mathrm{Pic}(R_P) = \mathbf{Z}/b\mathbf{Z}$. Of course e = d+1 in this case, so the torsion-free rank of $\mathrm{Pic}(R_P)$ is zero, and thus $\mathrm{Pic}(R_P)$ is a quotient of $\mathbf{Z}/a\mathbf{Z}$, the a arising in the expression for Cok(B). Let $\mathfrak{p}$ be the minimal prime order ideal corresponding to the facet of nK that does not contain the origin. Then $1/P_i$ belongs to $\mathfrak{p}^{n-n(i)}$ but not to any higher power of $\mathfrak{p}$, so $(1/P_i)R_P = \mathfrak{p}^{n-n(i)}$. These generate all of the additional relations on $\mathbf{Z}/a\mathbf{Z}$, so we deduce that the order of $[\mathfrak{p}]$ in $\mathrm{Pic}(R_P)$ is p = gcd{n−n(i)} = gcd{n(i)}. It is easy to see that $[\mathfrak{p}]$ generates $\mathrm{Pic}(R_P)$ in this case, and so $\mathrm{Pic}(R_P) = \mathbf{Z}/p\mathbf{Z}$.∎

Now we observe that the observation in 2 above, together with the result of Example A-1 can be used to show that generically, $R_P$ is almost never factorial. We require a result (more or less implicit in what has been done previously):

**A.8 PROPOSITION** If P is a projectively faithful element of $\mathbf{R}[x_i^{\pm 1}]$, then $R_P$ being regular entails that P be a chondrophage.

Proof: As $R_P$ is integrally closed, so is $R_S$, where S = Log P. Now $\underline{R}_P = R_P$, so that if Q belongs to $\mathbf{R}[x_i^{\pm 1}]^+$ with cvx Log Q = e·cvx Log P for some integer e, we have that $s = Q/P^e$ belongs to $R_P$. Thus $R_Q \subset R_P[s^{-1}] \subset R_{eK}$ (where as usual, K = cvx Log P). Hence $R_{eK}$ is integrally closed and regular, and so is factorial (as finitely generated projectives are free). So even at e = 1, K is integrally simple. Let v be a vertex of K; clearly v ∈ Log P. Let v' be a nearest neighbour to v along an edge of K. Then $x^{v'}/P$ lies in $R_P$ as the latter is integrally closed, and there thus exists an integer n with v' + nLog P ⊂ (n+1)Log P. If we put v = **0**, and $K_v$, to be the standard d-simplex with v' = (1,0,...,0,0) (as we may), then v' belonging to (n+1)Log P easily implies that v' belongs to Log P. Thus $K_v \subset$ Log P, so P is a chondrophage.∎

**A.8A THEOREM** Let P be an element of $\mathbf{R}[x_i^{\pm 1}]^+$ that is projectively faithful, and irreducible as an element of $\mathbf{R}[x_i^{\pm 1}]$. If $R_P$ is factorial, then up to the natural action of AGL(d,Z) on $\mathbf{Z}^d$, there exist positive real numbers $\{\lambda_i;\ i = 0, 1, ..., d\}$ such that

$$P = \lambda_0 + \sum_{i=1}^{d} \lambda_i x_i .$$

If additionally, $P$ belongs to $\mathbf{Z}[x_i^{\pm 1}]^+$, and $\tau_i$ are the pure states corresponding to the vertices of Log P, then

$$P = n(0) + \sum_{i=1}^{d} n(i)\,x_i ,$$

where $n(i)$ $(0 \leq i \leq d)$ are the unique positive integers such that $\tau_i(R_{P,\mathbf{Z}}) = \mathbf{Z}[1/n(i)]$.

**REMARK:** As being irreducible is generic (in the sense that a "random" polynomial will be irreducible, and we may always assume from the outset that our choice for $P$ is projectively faithful, this result says that $R_P$ being factorial is an extremely restrictive assumption; essentially the behaviour is that of Example 1A. Moreover, any choice of $\{\lambda_i\}$ will yield the same (i.e., order-isomorphic) $R_P$.

Proof: As $R_P$ is factorial, $P$ is a chondrophage (A8), and thus $K = \text{cvx Log } P$ is integrally simple. By statement 2 above, irreducibility of $R_P$ entails that $e = d+1$; thus $K$ is a simplex. By VIII.15, $K$ must be a multiple of the standard solid unit d-simplex (up to the action of $AGL(d,\mathbf{Z})$), say $K$ is $t$ times the standard simplex. From Example A-1, irreducibility of $P$ yields that the $a$ of statement 2 must equal $t$; thus $t = 1$. So $K$ is the standard unit simplex, and as $P$ is projectively faithful, Log P must be all of $K \cap \mathbf{Z}^d$. This yields the first conclusion.

If we now assume that $P$ has integer coefficients, then the $\lambda_i$ must all be integers, and it is straightforward to verify that in $R_{P,\mathbf{Z}} = \mathbf{Z}[x_i/P]$, the $\tau_i$ send $x_i/P$ to $1/\lambda_i$, and $x_j/P$ to 0, and the final statement follows.∎

It is possible to prove the preceding by different means (without going through integral simplicity); we outline an alternate proof:

By VIII.18, for all $w$ in Log P, $x^w/P$ is irreducible in $R_P$. If there were an affine relation among the elements of Log P, say of the form $\sum \alpha_i w_i = \sum \beta_j v_j$, with $\{w_i\}$ being disjoint from $\{v_j\}$, and $\sum \alpha_i = \sum \beta_j = 1$ with all the $\alpha$'s and $\beta$'s being positive, then there would exist an affine relation involving the same sets of lattice points with the $\alpha$'s and $\beta$'s all being positive and *rational* [H1; III.1A]. Hence there would exist a positive integer $N$ such that the $N\alpha_i$ and $N\beta_j$ are all integers. We would then obtain the relation:

$$\prod \left(\frac{x^{w_i}}{P}\right)^{N\alpha_i} = \prod \left(\frac{x^{v_j}}{P}\right)^{N\beta_j} .$$

However, if v and w are distinct lattice points in Log P, then $x^v/P$ and $x^w/P$ cannot be associates; their quotient, $x^{v-w}$ would have to belong to $R_P$, but it is not bounded on $(\mathbf{R}^d)^{++}$. Thus unique factorization has been contradicted. Hence, there can be no such affine relations. Thus Log P must be precisely the vertices of a simplex. As P is projectively faithful, it easily follows that K must be the standard simplex.∎

**EXAMPLE A-2.** Let K be the trapezoid with vertices $\{\mathbf{0}, (3,0), (1,1), (2,1)\}$ (illustration A1) and suppose that $\text{Log } P = K \cap \mathbf{Z}^2$. We calculate all possibilities for $\text{Pic}(R_P)$.

Here $K = K_1 + K_2$ where $K_1 = \text{cvx}\{\mathbf{0}, (2,0), (1,1)\}$ (so $K_1$ is *not* integrally simple, although K is), and $K_2 = \text{cvx}\{\mathbf{0}, (1,0)\}$. As $e = 4$, $\text{Cok}(B) = \mathbf{Z} \oplus \mathbf{Z}/a\mathbf{Z}$; the matrix B is

$$\begin{bmatrix} 1 & 0 & 0 \\ 0 & 1 & 0 \\ 0 & 0 & 1 \\ * & * & * \end{bmatrix}$$

so $a = 1$. There are now two possibilities. If P is irreducible, then $\text{Pic}(R_P) = \text{Cok}(B) = \mathbf{Z}$. Otherwise, $P = P_1P_2$, where $\text{cvx Log } P_i = K_i$. Let $\mathfrak{p}$ correspond to the right most edge (facet) of K; then $1/P_2 \in \mathfrak{p} \setminus \mathfrak{p}^2$, but $1/P_i$ belongs to no other minimal prime order ideal, so $\mathfrak{p} = (1/P_2)R_P$. Thus $\mathfrak{p}$ is principal, and from the form of B, it follows that $[\mathfrak{p}]$ generates $\text{Pic}(R_P)$. Hence $\text{Pic}(R_P)$ is zero, and so in this case, $R_P$ is factorial.

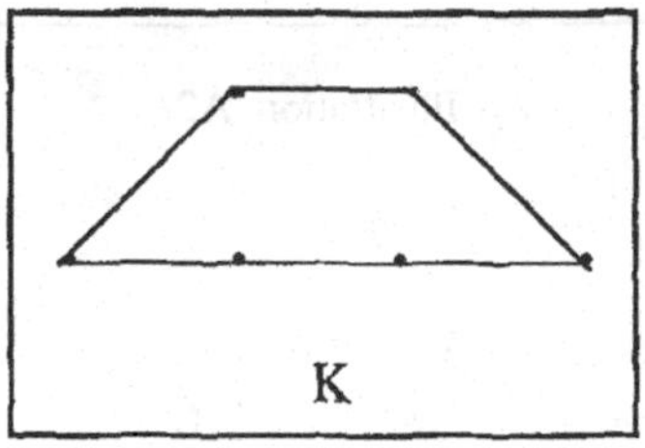


Illustration A1.∎

**EXAMPLE A-3.** Let P be a positive polynomial with

$$\text{Log } P = \{\mathbf{0}, (0,1), (1,0), (1,1), (1,2), (2,1), (2,2), (3,2)\}.$$

Then $K = \text{cvx Log } P$ is integrally simple and $K \cap \mathbf{Z}^2 = \text{Log } P$. We see that K has three indecomposable summands (all integrally simple), see illustration A2. All of the indecomposables are already in reduced form, so if P factors completely (i.e., into three irreducibles), $\text{Pic}(R_P) = 0$.

The matrix B has the form:
$$\begin{bmatrix} 1 & 0 & 1 \\ 0 & 1 & 0 \\ 0 & 0 & 1 \\ * & * & * \\ * & * & * \end{bmatrix}$$

where we have used the labelling of the faces of K as indicated in the illustration below. In particular, $\text{Cok}(B) = \mathbf{Z}^2$, and this is $\text{Pic}(R_P)$ if P is irreducible. If P neither factors completely, nor is irreducible, then we may write $P = P_1P_2$ with $\text{cvx Log } P_1 = K_i$, and $\text{cvx Log } P_2 = K_j + K_k$, where $\{i, j, k\} = \{1, 2, 3\}$. In this case, there is exactly exactly one new relation added to Cok(B), and it is easy to check that it is unimodular; thus $\text{Pic}(R_P) = \mathbf{Z}$.

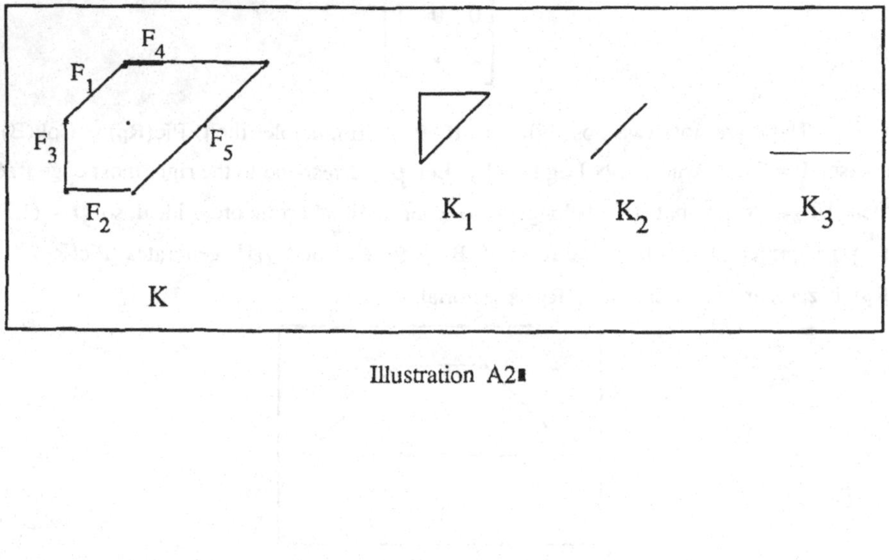


Illustration A2∎

## Appendix B. MORE ON THE FIXED POINT ALGEBRA.

Let $\beta : T^d \longrightarrow M_nC$ be a representation of the (d-)torus, having character $P$ in $\mathbf{Z}[x_i^{\pm 1}]^+$. Form the infinite tensor product algebra, $A = \otimes M_nC$, and the fixed point algebra

$$A^T = \{a \in A \mid a = \alpha(g)(a), \text{ for all } g \text{ in } T^d\}$$

where $\alpha = \otimes \mathrm{Ad}\beta : T^d \longrightarrow \mathrm{Aut}(A)$. Then $K_o(A^T)$ may be identified with the integer version of $R_P$, $R_{P,Z} = \mathbf{Z}[x^w/P \mid w \in \mathrm{Log}\, P\}$, with the ordering determined by the $\{x^w/P\}$ (see section I; also [H1; §I] and [HR1]). The order ideals of $K_o(A^T)$ are in bijection with the ideals of $A^T$ (and the closed ideals of the norm closure of $A^T$), and after the identification given below, these are generated by terms of the form $x^w/P^m$, for $w \in m\,\mathrm{Log}\, P$.

The identification is as follows. We may assume $\beta$ (and therefore $\overset{m}{\otimes}\beta$ ) is diagonal, say given by

$$t \mapsto \mathrm{diag}\,(t^{w_1}, t^{w_2}, \ldots, t^{w_n}),$$

for $t$ in $T^d$; of course $P = \sum x^{w_i}$ (with repetitions allowed). Define $e_i$ in $(M_nC)^\beta$ (the fixed point algebra under the action of $\beta$), to be the idempotent with a one in the (i,i) position and zeroes elsewhere. Then the class of $e_i$ in $K_o(A^T)$ corresponds to $x^{w_i}/P$; if $w$ is a lattice point expressible as a sum $\sum_{1 \le i \le m} w_{j(i)}$, with each $w_{j(i)}$ in $\mathrm{Log}\,P$, then the class of $e_{j(1)} \otimes e_{j(2)} \otimes \ldots \otimes e_{j(m)}$, (an element of $(\overset{m}{\otimes} M_nC^{\otimes\beta})$ in $K_o(A^T)$ corresponds to the product $x^w/P^m = \prod (x^{w_{j(i)}}/P)$.

The primitive ideals of $A^T$ correspond bijectively with the order ideals of $K_o(A^T)$ that are meet-irreducible. Order ideals of $K_o(A^T)$ are actually (ring) ideals in the latter [H1; I.2]. The result in II.7 indicates that the localization obtained by inverting the order units induces a bijection on the order ideals, preserving notions such as meet- irreducibility. This can be exploited as follows.

Let $P = 1 + x_1 + \ldots + x_d$, $P' = \lambda_o + \sum \lambda_i x_i + \sum \lambda_{jk} x_j x_k$, with the $\lambda$'s all being positive integers. Then $P$ corresponds to the standard representation $t \mapsto \mathrm{diag}(1, t_1, \ldots, t_d)$, where $t = (t_1, \ldots, t_d) \in T^d$, and $P'$ is the character of a representation $\beta'$ of dimension $n = \lambda_o + \sum \lambda_i + \sum \lambda_{jk}$. Observe that $\mathrm{Log}\, P' = \mathrm{Log}\, P^2$ so that $R_{P,Z}$ and $R_{P',Z}$ have the same localization (obtained by inverting the order units), and this is $R_{K,Z}$ (where we have used rational rather than real coefficients in the latter's definition) with $K$ the standard simplex. Now $R_{P,Z}$ is a pure polynomial ring, while $R_{P',Z}$ is virtually never even factorial (Theorem A8A). However, because $K$ is integrally simple, we can write down generators for all the primitive ideals in the fixed point algebra of $A = \otimes M_n C$ under $\alpha' = \otimes \mathrm{Ad}\beta'$.

For each facet F of K, set $z_F = \sum_{w \in F \cap Z^d} x^w/P'$. Then $z_F$ corresponds to a sum of orthogonal, diagonal idempotents in $(M_n C)^{\beta'}$, $e_F$, and this is of course idempotent. The set of generators described in section VII that yield the meet-irreducible order ideals in $R_{K,Z}$, $\{z_{F_1}^{n(1)}, z_{F_2}^{n(2)}, \ldots\}$ translate to the set of tensor product idempotents

$$\{e_{F_1} \otimes e_{F_1} \otimes \ldots \otimes e_{F_1}, e_{F_2} \otimes e_{F_2} \otimes \ldots \otimes e_{F_2}, \ldots\}$$

(the first term is the n(1)-fold tensor product of copies of $e_{F_1}$, etc.). These generate (as a two-sided ideal) the corresponding primitive ideal of the fixed point algebra, $A^T$. The primitive subquotients of $A^T/I$ (where I is one of these primitive ideals) can be determined using dimension group theory; for example, if the order ideal is contained is just one maximal ideal (which is equivalent to the intersection of the $F_i$'s being a singleton, i.e., a vertex), then $A^T/I$ is an iterated extension of algebras of compact operators tensored with an $m^\infty$ UHF algebra for some $m < n$ (including the possibility that $m = 1$ occur— see [H1; § VII]).

We have frequently used the fact proved in [H1; § VII] that the order ideals that are prime in $R_P$ (and thus correspond to similar order ideals in $R_K$) are obtained from the following prescription. Let F be a face of $K = \text{cvx Log } P$, let $\mathfrak{p}_F$ be the ideal generated by $\{x^w/P \mid w \in \text{Log } P \backslash F\}$. Then $\mathfrak{p}_F$ is an order ideal by II.2A, and it is the kernel of the following map.

For $f = \sum \lambda_v x^v$ in $\mathbf{R}[x_i^{\pm 1}]$ with $\text{Log } f \subset n \text{Log } P$, set

$$f_{nF} = \sum_{v \in nF \cap Z^d} \lambda_v x^v .$$

Form $R_{\mathfrak{p}_F}$; we can now define a mapping,

$$\pi_F : R_P \to R_{P_F}$$
$$f/P^n \mapsto f_{nF}/P_F^n ;$$

this is a well-defined ring homomorphism, it is order-preserving, and an order-epimorphism ($\pi_F(R_P^+) = R_{P_F}^+$). It follows that $R_P/\mathfrak{p}_F \cong R_{P_F}$ as ordered rings. Finally, we have also used the fact that every order ideal of $R_P$ that is prime is of the form $I = \mathfrak{p}_F$ for some face F of K.

We shall give an alternate proof of these results (except the last sentence) using the fixed point algebra $A^T$ that was described earlier. Now this requires that we work with polynomials with integer coefficients, and the corresponding version of $R_K$, $R_{K,Z}$, uses rational coefficients as opposed to real ones. However, the "real" version of this result can be deduced from the integer form without much difficulty.

Owing to the natural bijection between order ideals of $R_P$ and ideals of $A^T$ (assuming that P belongs to $\mathbf{Z}[x_i^{\pm 1}]^+$), the $\mathbf{p}_F$'s correspond to a very special class of ideals of $A^T$; let $J_F$ be the corresponding two-sided ideal of $A^T$, then the quotient $A^T/J_F$ will turn out to be a fixed point algebra under a xerox type action on a smaller version of A, with a character corresponding to $P_F$. We explicitly exhibit the onto homomorphism of algebras, $A^T \longrightarrow A_o^{T_o}$, where $T_o$ is a torus of lower dimension, (possibly zero-dimensional) with kernel $J_F$, and so provide the desired proof.

So let $\beta: T^d \longrightarrow M_n C$ be a diagonal representation implementing the character P, form the j-fold tensor product of copies of $M_n C$, $A^j$, and the corresponding j-fold tensor action $\otimes\beta$ on $A^j$. Let $(A^j)^T$ denote the corresponding fixed point algebra; this is of course, the centralizer of $\overset{j}{\otimes}\beta(T^d)$ in $A^j$. The maps $(A^j)^T \longrightarrow (A^{j+1})^T$ are obtained via the inclusions $A^j \longrightarrow A^{j+1} = A^1 \otimes A^j$, given as $a \mapsto I_n \otimes a$. Then $A = \lim A^j$ and $A^T = \lim (A^j)^T$.

Write, for t in $T^d$, $\beta(t) = \mathrm{diag}\,(t^{w_1}, t^{w_2}, \ldots, t^{w_n})$, where $w_i \in \mathbf{Z}^d = (T^d)^\wedge$. Then define P $= \sum x^{w_i}$, with repetitions of course allowed. Form $K = \mathrm{cvx}\,\mathrm{Log}\,P$.

Let F be a face of K, and let E be the diagonal idempotent in $M_n C$ given by $E_{ii} = 1$ if and only if $w_i$ belongs to F. Then E lies in the centre of $(M_n C)^T$ as follows immediately from the definitions. Define the j-fold tensors, $E^j = E \otimes \ldots \otimes E$ in $A^j$; the $E^j$ arise from the j-th power of P, $P^j$ (with the face jF of jK playing the rôle of F in K) in the same way that E did from F, and so $E^j$ lies in the centre of $(A^j)^T$. Let $B = EM_n CE$, and form the j-fold tensor product algebras $B^j = \overset{j}{\otimes} B$, and the limit algebra $A_o = \lim B^j$. Define a representation $\beta_0$ of $T^d$, $\beta_0: T^d \longrightarrow B$, via $\beta_0(t) = \mathrm{diag}(\ldots, t^{w_i}, \ldots)$ where the i runs *only over indices for which $w_i$ belongs to F*. We obtain corresponding fixed point algebras $(B^j)^T$, and its limit as j increases, $A_o^T$. The character of $\beta_0$ is of course $P_F$.

We claim that $\psi(a) = \lim E^j a E^j$ yields a well-defined expectation $\psi: A \longrightarrow A_o$, and that its restriction to $A^T$ is an algebra homomorphism onto $A_o$. Just observe that $E^{j+1} = E \otimes E^j$, and that $A^j \longrightarrow E^j A^j E^j = B^j$ is onto and completely positive. For a and b in $(A^j)^T$, $E^j ab E^j = E^j a E^j E^j b E^j$ as a and b commute with $E^j$; it easily follows that $\Psi = \psi|A^T$ is an algebra homomorphism (of course, $\{E^j a E^j\}_{j\to\infty}$ is eventually stationary for each a in $\bigcup_j ((A^j)^T)$ ). Clearly $E^j (A^j)^T E^j = (B^j)^T$ , so that $\Psi$ is an algebra homomorphism onto $A_o^T$.

The kernel of $\Psi$ at the j-th level is $(1-E^j)(A^j)^T$; a complete set of equivalence classes (in the relevant $K_o$ group, within the set of classes between [0] and [1]) is given by the diagonal idempotents, which of course correspond to $x^w/P^j$, where w belongs to jLog P. Since every ideal of $(A^j)^T$ is generated as a two-sided ideal by idempotents which generate the kernel of the corresponding quotient map on $K_o$, we obtain that $\ker K_o(\Psi)$ is

$$\sum_{w \in j \operatorname{Log} P \setminus jF} \left( \frac{x^w}{P^j} \right) \mathbf{Z}.$$

The map induced on $K_0$ by $\Psi$ sends $x^w/P^j$ to zero if $w \notin jF$, and to $x^w/P_F^j$ if $w \in jF$; this follows from the identification of idempotents in $A_0^T$ with elements of $R_{P_F}$. As $A^T$ is von Neumann regular (since we have no need to close up any of these algebras), $\Psi$ being onto implies that $K_0(\Psi)$ is as well. Verification of the rest of the details is completely straightforward.

The proof of the fact that every prime order ideal of $R_P$ is of the form $\mathfrak{p}_F$ for some facet F of K is rather involved, as it requires the material in [H1; § III and VII]. I could not find a proof using the algebra homomorphism; difficulties arise as it is the *meet-irreducible* order ideals (not just prime ones) that correspond to primitive ideals of $A^T$.

## Appendix C. VARIABLE POLYNOMIALS

In this section, we consider a variation on the original positive polynomial problem; instead of simply multiplying our original target polynomial f by the *same* polynomial P repeatedly, we allow the "P" to vary somewhat. The main theorem of this appendix (C1) asserts that modest variations in the coefficients of the P's does not change the solution to the positivity problem. A surprising corollary (C2) which applies to the original situation, comes out of the argument.

**C1 THEOREM** Let $\{P_i\}_{i=1}^{\infty}$ be a sequence of elements in $\mathbf{R}[x_i^{\pm 1}]^+$, let P be another element in $\mathbf{R}[x_i^{\pm 1}]^+$, and let f belong to $\mathbf{R}[x_i^{\pm 1}]$. Suppose there exists an infinite subset J of $\mathbf{N}$ such that:

(i) For all j in J, $\operatorname{Log} P_j = \operatorname{Log} P$.

(ii) For each monomial $x^w$ appearing (with nonzero coefficient) in P, the coefficients of $x^w$ in $\{P_j/P_j(1)\}_{j \in J}$ are bounded below.

If for some n, $P^n f$ has no negative coefficients, then there exists a finite subset J' of J such that

$$\left(\prod_{j \in J'} P_j\right) \cdot f \qquad \text{has no negative coefficients.}$$

To prove this, we may assume $J = \mathbf{N}$. Define a norm on $\mathbf{R}[x_i^{\pm 1}]$ via $\|g\| = \max\{|\mu_w| \mid w \in \operatorname{Log} g\}$, where $g = \sum \mu_w x^w$ (so the norm is that of the Fourier transform of g, g^, as an element of $l^\infty(\mathbf{Z}^d)$. Define the following subsets of $\mathbf{R}^{\#(\operatorname{Log} P)}$:

$$M_P = \{Q \in \mathbf{R}[x_i^{\pm 1}]^+ \mid \operatorname{Log} Q = \operatorname{Log} P\}$$

$$L_{P,t} = \{Q \in M_P \mid \operatorname{Log} Q^t f = t \operatorname{Log} Q + \operatorname{Log} f\}; \text{ here } t = 1, 2, \ldots .$$

Note that $\operatorname{Log} Q^t f \subset t \operatorname{Log} Q + \operatorname{Log} f$ (according to our convention on sums of sets, the right side is the set of sums of t elements of Log Q with one from Log f) in general, and as Q has only positive coefficients $\operatorname{Log} Q^t = t \operatorname{Log} Q$. Clearly $M_P$ is an open subset of $\mathbf{R}^{\#(\operatorname{Log} P)}$. We show that for every integer value of t, $M_P \setminus L_{P,t}$ is contained in a closed set of measure zero.

To this end, write $f = \sum \mu_w x^w$ and observe that for $t = 1$, if $Q = \sum \lambda_v x^v$ belongs to $M_P \setminus L_{P,t}$, then there exists u in $\operatorname{Log} Q + \operatorname{Log} f$ such that $\sum \lambda_v \mu_w = 0$, where the sum is over $\{(v,w) \mid v \in \operatorname{Log} Q,\ w \in \operatorname{Log} f, \text{ and } v + w = u\}$. In other words, the coefficients $\{\lambda_v\}$ must lie in the hyperplane defined by a subset of $\{\mu_w\}$. As $\operatorname{Log} Q + \operatorname{Log} f$ is finite, $M_P \setminus L_{P,1}$ is contained in a finite union of hyperplanes, so its closure is of zero measure and nowhere dense.

For $t > 1$, the coefficients of $Q^t$ are polynomials in the $\lambda_v$'s, so for $Q$ not to belong to $L_{P,t}$ the $\lambda_v$'s would have satisfy one of a finite number of polynomials (one arising from each member of $\operatorname{Log} Q + \operatorname{Log} f$; so again, $M_P \setminus L_{P,t}$ is contained in a closed set of measure zero.

Thus $\bigcap_{t=1}^{\infty} L_{P,t}$ is dense in $M_P$. Select $Q$ in this intersection; that is, $Q$ has no negative coefficients, $\operatorname{Log} Q = \operatorname{Log} P$ and and for all $t$, $\operatorname{Log} Q^t f = t\operatorname{Log} Q + \operatorname{Log} f$. We use the density of the intersection only to obtain such a $Q$.

Equipped with the norm defined earlier, $\{P_j/P_j(1)\}$ is a bounded sequence in the finite dimensional space $M_P$, and so contains a limit point, $P_0$. We may thus find a subsequence $\{P'_1, P'_2, \ldots\}$ of $\{P_j/P_j(1)\}$ that converges to $P_0$ in norm. This means that for every $w$ in $\operatorname{Log} P$, the sequence {coefficient of $x^w$ in $P'_j$} converges to the coefficient of $x^w$ in $P_0$. We deduce from the boundedness-below hypothesis that $\operatorname{Log} P = \operatorname{Log} P_n$.

Applying Theorem II.1 twice, there exists an integer $m$ so that both $Q^m f$ and $P_0^m f$ have no negative coefficients. Now form the partially ordered rings, $R_0 = R_{P_0}$, $R_Q$, and $R_{P_0Q}$ corresponding to the polynomials $P_0$, $Q$, and $P_0 Q$ respectively. By the argument at the conclusion of the proof of II.1, we may assume there exists an integer $e$ so that $\operatorname{Log} f \subset e\operatorname{Log} Q$. Hence we may define the elements $a = f/P_0^e$ in $R_0 \subset R_{P_0Q}$, and $a' = f/Q^e$ in $R_Q \subset R_{P_0Q}$. As $P_0^m f$ and $Q^m f$ have no negative coefficients, $a$ and $a'$ are both positive elements in the relevant partially ordered rings.

We claim that for each $w$ in $\operatorname{Log} f$, there exists a positive real number $d$ (depending on $w$) such that $a' \geq d(x^w/Q^e)$, the inequality holding with respect to the partial ordering in $R_Q$. Notice that $w$ belongs to $\operatorname{Log} f$, and this is contained in $e\operatorname{Log} Q$, so $x^w/Q^e$ belongs to $R_Q$, and also to $R_P$. As $Q^m f$ has no negative coefficients, we have that the coefficient of $x^{w+u}$ in $Q^m f$ is non-negative, for any $u$ in $m\operatorname{Log} Q$. As $Q$ is a member of $\bigcap_{t=1}^{\infty} L_{P,t}$, $w + m\operatorname{Log} Q \subset \operatorname{Log} Q^m f$; hence the coefficients of the $x^{w+u}$ are all nonzero, and are thus strictly positive. Therefore there exists an integer $N$ so that the coefficient of any monomial in $NQ^m f$ exceeds its coefficient in $x^w Q^m$. Dividing by $NQ^{m+e}$, we obtain $a' \geq d(x^w/Q^e)$, in $R_Q$ (where $d = 1/N$).

The element $Q^e/P_0^e$ is an order unit of $R_{P_0Q}$ [H1; V.5], so by II.5, $a \geq d(x^w/Q^e)$ in $R_{P_0Q}$, and by II.1, this latter inequality holds in $R_{P_0}$.

By applying this process successively to $a$, $a - d_1(x^{w_1}/P_0^e)$, $a - d_1(x^{w_1}/P_0^e) - d_2(x^{w_2}/P_0^e)$ ($d_2$ depending on $d_1$ and $w_2$), etc., running over all the $w_i$ in $\operatorname{Log} f$, we find there exists a positive real number $D$ so that for all real numbers $D_w$ with $0 < D_w < D$,

$$a \geq \sum_{w \in \operatorname{Log} f} D_w \cdot \left( \frac{x^w}{P_o^e} \right) \quad \text{in } R_{P_o}.$$

Translating this back to polynomial multiplication, there exists a positive real number D so that for all g in $\mathbf{R}[x_i^{\pm 1}]^+$ with $\operatorname{Log} g \subset \operatorname{Log} f$ and $\|g\| < D$, there exists M (depending on g) so that $P_o^e \cdot (f-g)$ has no negative coefficients.

An immediate corollary is the following, which says that if $P^n f$ has non-negative coefficients, eventually (i.e., for all sufficiently large powers of P) all monomials that could possibly occur with nonzero coefficient in $P^N f$, actually do, even though some cancellation of terms might have occurred:

**C2 COROLLARY** If P is a polynomial in $\mathbf{R}[x_i^{\pm 1}]$ with no negative coefficients, and f belongs to $\mathbf{R}[x_i^{\pm 1}]$, and for some integer n, $P^n f$ has only non-negative coefficients, there exists an integer s so that for all $t \geq s$, $\operatorname{Log} P^t f = t \operatorname{Log} P + \operatorname{Log} f$.

Proof: One inclusion is automatic. By the result immediately above, there exists a positive real number D so that if g in $\mathbf{R}[x_i^{\pm 1}]^+$ satisfies $\|g\| < D$ and $\operatorname{Log} g \subset \operatorname{Log} f$, then there exists M so that $P^M \cdot (f-g)$ has no negative coefficients, and of course the same is true if M is replace by any larger integer. For t exceeding M and n, the coefficients of terms in $P^t f$ dominate their counterparts in $P^t g$, so that $\operatorname{Log} P^t g \subset \operatorname{Log} P^t f$. However, g has only positive coefficients, so $\operatorname{Log} P^t g = \operatorname{Log} P + \operatorname{Log} g$. As we may assume $\operatorname{Log} g = \operatorname{Log} f$, we obtain the desired inclusion.∎

Finally, we can conclude the proof of the theorem. We may apply C2 and II.1 to our situation; there exists an integer t such that $\operatorname{Log} P_o^t f = t \operatorname{Log} P_o + \operatorname{Log} f$. Let E be the infimum of the nonzero coefficients of $P_o^t f$. There exists a finite subset $\{P_{s(1)}, P_{s(2)}, \ldots, P_{s(t)}\}$ of $\{P_j/P_j(1)\}$ so that

$$\left\| \prod_{i=1}^{t} P_{s(i)} - P_o^t \right\| < \frac{E}{\|f\|}.$$

Thus $\|(\prod P_{s(i)} - P_o^t) \cdot f\| < E$. Hence if $x^w$ appears in $P_o^t f$ (with nonzero coefficient), it appears in $(\prod P_{s(i)}) \cdot f$ with positive coefficient. However, $\operatorname{Log} P_{s(i)} = \operatorname{Log} P$, so $\operatorname{Log}\big((\prod P_{s(i)}) \cdot f\big)$ is contained in $t \operatorname{Log} P + \operatorname{Log} f = \operatorname{Log} P_o^t f$. Thus all monomials that appear in $(\prod P_{s(i)}) \cdot f$ actually appear in $P_o^t f$, and therefore have positive coefficients.∎

The approximation portion of the argument is of course trivial—difficulties arose because of the probability zero possibility that cancellation would occur in $P_0^t f$.

If we are dealing with one variable, say $x$, and if $P_i = 1 + a_i x$ (where $a_i$ are positive real numbers), then the hypotheses of the theorem will be satisfied if there exists $N$ so that $\{i \mid |\mathrm{Log}\, a_i| < N\}$ is infinite (with $P = 1 + x$, and $f$ in $\mathbf{R}[x^{\pm 1}]$ satisfying $f|(0,\infty) > 0$). If instead, $a_i = i^2$, then the conclusion of the theorem actually fails. To see this, we note that in this case, the infinite product $\prod (P_i/P_i(1))$ converges to an entire function, say $h$, with Taylor series $b_0 + b_1 x + \ldots$; if $\alpha$ and $\beta$ are real numbers chosen in order that $\alpha > b_1/b_0$ and $\alpha^2 < 4\beta$, then $f = \beta x^2 - \alpha x + 1$ can be made positive by repeated multiplication by $1 + x$, but the $x$ coefficient of $h \cdot f$ will be negative, and it follows easily that the conclusion of the theorem fails. If $\alpha_i = i$, recent work by B.M. Baker and the author [BH] has shown that in fact the conclusion of the theorem succeeds.

There is a converse to the theorem: If the $P_i$ all satisfy $\mathrm{Log}\, P_i = \mathrm{Log}\, P$, and if there exists a finite set $J_0$ such that $(\prod_{i \in J_0} P_i) \cdot f$ has no negative coefficients, then there exists an integer $n$ so that $P^n f$ also has no negative coefficients (notice that we can forget about the bounded below hypotheses). To see this, set $Q = \prod_{i \in J_0} P_i$, and observe that $\mathrm{Log}\, Q = m \mathrm{Log}\, P$ (where $m$ is the cardinality of $J_0$), and therefore, we can apply II.1 to $Q$ and $P^m$.

Since we did not use the full force of II.1 in the preceding arguments, it should be possible to weaken the hypotheses from $\mathrm{Log}\, P_i = \mathrm{Log}\, P$ (in the statement of the Theorem) to $\mathrm{cvx}\, \mathrm{Log}\, P_i = \mathrm{cvx}\, \mathrm{Log}\, P$ (keeping the bounded below condition on the coefficients of the monomials appearing in $P_i$). But now the possibility of cancellation, arising from the extra terms, appears as a possible obstruction.

Using techniques from [BH] and those developed here, it is possible to prove results of the following sort:

> Let $P_i$ be in $\mathbf{R}[x_i^{\pm 1}]^+$ and satisfy $\mathrm{Log}\, P_i = (iK) \cap \mathbf{Z}^d$ for some fixed integral polytope $K$, and suppose that the nonzero coefficients are bounded below. Let $Q$ in $\mathbf{R}[x_i^{\pm 1}]^+$ be such that $\mathrm{Log}\, Q = (dK) \cap \mathbf{Z}^d$. If for $f$ in $\mathbf{R}[x_i^{\pm 1}]$, there exists an integer $m$ such that $Q^m f$ has no negative coefficients, then given $k$, there exists $n$ exceeding $k$ such that $P_k P_{k+1} \ldots P_n f$ has no negative coefficients.

## Appendix D. INTEGRALLY SIMPLE POLYTOPES ARISING FROM REFLECTION GROUPS

Let P in $\mathbf{Z}[x_i^{\pm 1}]^+$ $(i \leq d)$ be the restriction of an irreducible character of one of the compact connected semisimple Lie groups of rank d. We show that for almost all choices of P (specifically, those whose dominant weight lies in the interior of the Weyl chamber), $R_P$ is homologically regular. This we do by showing that after adjusting P in order to be projectively faithful, the relevant convex hull is an integrally simple polytope. Each integral polytope that results in this fashion is the convex hull of a single orbit of the Weyl group acting on $\mathbf{Z}^d$, but integral simplicity is determined with respect to a (generally) smaller sublattice of $\mathbf{Z}^d$. When $d = 2$, this result is extended to the actions of a few other reflection groups, the dihedral groups of index 6 and 4 respectively.

For our purposes, we define a lattice in $\mathbf{R}^d$ to be a subset L of $\mathbf{Z}^d$ of the form $L_o + z$, where L is a subgroup of $\mathbf{Z}^d$ and z is some point of $\mathbf{Z}^d$. A compact convex polytope $K \subset \mathbf{R}^d$ is called L-integral (for a lattice L) if the vertices of K lie in L. The L-integral polytope K is called L-integrally simple if both of the following hold:

(i) K is simple as a real polytope.

(ii) For every vertex v of K, the convex hull of the nearest points to v in $L \cap K$ along edges of K, together with v, is a fundamental simplex for L.

An equivalent way to express (ii) if $\mathbf{Z}^d/L_o$ (recall that $L_o$ is the subgroup of $\mathbf{Z}^d$ such that $L = L_o + z$) is finite, is:

> The volume of the peak polytope (computed with respect to L) described in (ii) is exactly $|\mathbf{Z}^d/L_o|/d!$.

Let W be the Weyl group of some compact connected semisimple Lie group G of rank d. Then W acts on the dual of the maximal torus, i.e., on $\mathbf{Z}^d$, as a group of matrices in GL(d,Z). Let a in $\mathbf{Z}^d$ be a dominant weight with respect to the action of W. Define $K_a = \text{cvx}\{a^g \mid g \in W\}$. Let $P_a$ be the restriction to the maximal torus of the unique irreducible character of G whose dominant weight is a. Then if $L_o$ is the subgroup of $\mathbf{Z}^d$ generated by $\text{Log } P_a - \text{Log } P_a$ and $L = a + L_o$, it is well-known that $K_a \cap L = \text{Log } P_a$.

We shall show in particular, that if a lies in the interior of the Weyl chamber, then $K_a$ is L-integrally simple; this also holds for *some* values of a on the boundary, (but not all if d exceeds 2, or even if $d = 2$ and $G = SO(5)$).

The reason for considering L-integral simplicity is that $P_a$ is not projectively faithful (unless the centre of G is trivial), so that in order to apply the results of section VI to $R_{P_a}$, we would have to go down to the subgroup $L_o$ as defined above, take a basis for this, multiply $P_a$ by $x^{-a}$, and re-express the exponentials appearing in $x^{-a}P_a$ in terms of this basis. Then integral simplicity with respect to this basis is just what L-integral simplicity boils down to. Insofar as $R_{P_a}$ is independent of the choice of process (up to order-isomorphism), it seems easier to introduce and use L-integral simplicity.

When G and W are replaced by a finite reflection group acting on $\mathbf{Z}^d$, there is a choice that has to be made in the definition. We shall discuss this after dealing with the original situation.

I had proved the following by rather computational means for all of the essentially simple groups except $F_4$, $E_6$, $E_7$, and $E_8$. However, a discussion with Professor S. Berman (University of Saskatchewan) lead to the following very elementary (and more general) proof:

**D1 THEOREM** Let $a = (a_1, \ldots, a_d)$ in $\mathbf{Z}^d$ be a dominant weight for the Weyl group W of a compact connected semisimple Lie group G, and suppose that a lies in the interior of the Weyl chamber. Then $K_a$, the convex hull (in $\mathbf{R}^d$) of the orbit of a under W, is L-integrally simple.

**REMARK.** Actually, a little more is true if (S,W) is a Coxeter group (notation from N.Bourbaki, Groupes et Algèbres de Lie, Chap.4-6) so that the corresponding representation restricts to a representation over the integers [ibid; pp. 91-2[), then the same conclusion holds (same proof). However, notice that dihedral groups are Coxeter groups, their standard representations are 2-dimensional, yet if half the order is not one of 3, 4, or 6, then the representation does not so restrict.

Proof: Let $\{\lambda_1, \ldots, \lambda_d\}$ be a set of fundamental dominant weights for G (we are using notation from Introduction to Lie Algebras and Representation Theory, by J.E. Humphreys). We may write $a = \sum m_i \lambda_i$ where the $m_i$ are non-negative integers; because a is not on the boundary of the Weyl chamber, each $m_i$ is strictly positive. It follows that the lattice L is $L_a = a + \sum \lambda_i \mathbf{Z}$.

For each i, define $a^i = a - \lambda_i$; each of these in still a dominant weight and is also a weight of the irreducible character of G corresponding to a. Thus $a_i$ belongs to $K_a \cap L_a$. We claim that the $a_i$ are nearest neighbours in $L_a$ to a, along edges of $K_a$. Notice that $-\mathrm{cvx}\{a_i - a, a - a\} = -\mathrm{cvx}\{-\lambda_i, \mathbf{0}\}$ is the standard simplex for $L_o = \sum \lambda_i \mathbf{Z}$. Any other dominant weights that occur as weight for the irreducible character corresponding to a can be written in the form $\sum m_i' \lambda_i$, where $m_i' \leq m_i$. It follows immediately that $K_a$ is L-integrally simple at a. As W acts transitively on the vertices of $K_a$ (which are the images of a under the action of W), and the action is implemented within GL(d,Z) and leaves L stable, $K_a$ is L-integrally simple at all of its vertices, and so is L-integrally simple.∎

Characters "on the walls" may or may not yield L-integrally simple polytopes. For example, if $G = SU(4)$ and $a = (a_1, a_1, 0)$ with $a_1 > 0$, then $K_a$ is an octahedron; if $a = (a_1, a_2, a_2)$ with $a_1 > a_2 > 0$, then $K_a$ is an eicosahedron. In the former case, each vertex has 4 edges emanating from it, and in the lattter, 5. So in each of these cases, $K_a$ is not even simple (as a real polytope), let alone integrally simple. On the other hand, if $a = (1,0,0)$, corresponding to the standard character, then $K_a$ is the basic $L_a$-simplex, and is L-integrally simple. For $SO(n)$, $n \geq 5$, the "standard" character does not yield an L-integrally simple polytope; it is simple (as a real polytope) for $n = 5$, but not for $n > 5$.

For $G = SU(d+1)$, there is a complete characterization of the dominant weights $a$ such that $K_a$ is L-integrally simple:

**D2 PROPOSITION** If $G = SU(d+1)$ and $a = (a_1, \ldots, a_d)$ belongs to $\mathbf{Z}^d$ with $a_1 \geq a_2 \geq \ldots \geq a_d \geq 0$, then $K_a$ is L-integrally simple if and only if one of the following holds:

(i) $a_1 > a_2 > \ldots > a_d > 0$ ($a$ lies in the interior of the Weyl chamber);

or

(ii) there exists $k$ in $\{0, 1, \ldots, d\}$ so that $0 = a_d = a_{d-1} = \ldots = a_{k+1} < a_k < \ldots$ (define $a_0 = \infty$ if necessary);

or

(iii) there exists $k$ so that $a_1 = a_2 = \ldots = a_k > a_{k+1} > \ldots > 0$.

Furthermore, if none of (i), (ii), or (iii) hold, then $K_a$ fails even to be simple as a real polytope.

<u>Proof:</u> In the case that $G = SU(d+1)$, $W = S_{d+1}$, and the action is on $\mathbf{Z}^d$. Specifically, let $S_{d+1}$ act by permuting the basis elements of $\mathbf{Z}^{d+1}$, and identify $\mathbf{Z}^d$ with $\mathbf{Z}^{d+1}/(1, 1, \ldots, 1)$. In other words, if $\{e_1, \ldots, e_d\}$ is the standard basis for $\mathbf{Z}^d$, then $W$ acts by arbitrary permutations on the set

$$\{e_1, \ldots, e_d, -\textstyle\sum e_i\},$$

and this extends to an action on $\mathbf{Z}^d$.

Let $a$ be a point in $\mathbf{Z}^d$; the orbit of $a$ under $W$ contains a unique point (called a dominant weight), $a' = (a_1, \ldots, a_d)$ where $a_1 \geq a_2 \geq \ldots \geq a_d \geq 0$. So to compute $K_a$, we may assume that $a$ is already dominant. Then $a$ corresponds uniquely to an irreducible character $\chi_a$ of $SU(d+1)$, and the set of weights (lattice points) that appear in $P_a = \chi_a | T^d$ is precisely:

$$\left\{z = (z_1, \ldots, z_d) \in \mathbf{Z}^d \mid \textstyle\sum z_i \equiv \sum a_i \bmod (d+1)\right\} \cap \operatorname{cvx}\left\{a^g \mid g \in W\right\}.$$

Thus if $L = \{z \in \mathbf{Z}^d \mid \sum z_i \equiv \sum a_i \bmod(d+1)\}$, then $L_o = L-a$ is a subgroup of index $d+1$. Hence it is sufficient to show that the peak polytopes of $K_a$ have volume $(d+1)/d!$ Since the set of vertices of $K_a$ is precisely the orbit of a, it is sufficient to show that the peak polytope at a has the desired volume.

We consider those a satisfying (ii) to begin with. That is,

$$0 = a_d = a_{d-1} = \ldots = a_{k+1} < a_k < \ldots .$$

Subtract a from $K_a$, and consider the following points:

$$\begin{aligned}
e^1 &= (-1, 1, 0, \ldots, 0, 0, 0) \\
e^2 &= (0, -1, 1, 0, \ldots, 0, 0) \\
&\vdots \\
e^k &= (0, 0, \ldots, 0, -1, 1, 0, \ldots, 0, 0) \\
e^{k+1} &= (0, 0, \ldots, 0, -1, 0, 1, \ldots, 0, 0) \\
&\vdots \\
e^{d-1} &= (0, 0, \ldots, 0, -1, 0, \ldots, 0, 1) \\
e^d &= (-1, -1, \ldots, -1, -2, -1, \ldots, -1, \ldots, -1).
\end{aligned}$$

(Recall that $\{e_1, \ldots, e_d\}$ is the standard basis.)

We shall show: (a) $e^j$ belongs to $(K_a - a) \cap (L-a)$;

(b) given g in $S_{d+1}$, there exist non-negative integers $\lambda_j$ such that $a^g - a = \sum \lambda_j e^j$;

(c) $K_a - a$ is integrally simple relative to the lattice $L_o = L - a$.

To prove (a), simply note that for $j < k+1$. $(a_j - a_{j+1})e^j = a^g - a$, where g is the transposition interchanging $e_j$ and $e_{j+1}$. If $d \geq j \geq k+1$, let g be the transposition interchanging $e_k$ and $e_{j+1}$. Finally, $e^d$ is obtained by choosing g to interchange $e_k$ and $(-1, \ldots, -1)$, and then dividing $a^g - a$ by $a_k$.

Now to prove (b), first assume that g merely permutes $\{e_1, \ldots, e_d\}$ (leaving $e_{d+1} = (-1, \ldots, -1)$ fixed). Then if $a^g - a = (b_1, \ldots, b_d)$, we must have that $b_1 \leq 0$, $b_1 + b_2 \leq 0$,

$\sum_{1\le i\le j} b_j \le 0$ for $j = 1, \ldots, d$, and $\sum_{1\le i\le d} b_j = 0$; moreover, if $d > k$, then $b_{k+1}, \ldots, b_d$ must all be non-positive. Then we have:

$$a^g - a = -b_1 e^1 - (b_1 + b_2) e^2 - \ldots - \left(\sum_{i=1}^{k-1} b_i\right) e^{k-1} - b_{k+1} e^k - \ldots - b_d e^{d-1}$$

(If $k = d$, ignore the $e^k$ and subsequent terms).

Next, if $g$ is an arbitrary element of $S_{d+1}$, then $a^g - a = a^h - a - a_i(1,1,\ldots,1) - a_i e_{h(i)}$ for some $i \le k$, where $h$ is a permutation of $\{e_1, \ldots, e_d\}$. If $j = h(i) < k$, we may write $a^g - a = a^h - a + a_i(e^d + e^j + \ldots + e^{k-1})$; if $h(i) = k$, then $a^g - a = a^h - a + a_i e^d$; in these cases, we are done as $a^h - a$ is already a positive integer combination of $\{e_1, \ldots, e_{d-1}\}$. This leaves the situation that $d \ge j = h(i) > k$. As $a_i$ is not zero only for $i \le k$, we must have that $h$ is not the identity, and

$$\begin{aligned} a^g - a &= a^h - a + a_i(-1, \ldots, -1, -2, -1, \ldots, -1) \\ &= a^h - a + a_i e^d + a_i(0, 0, \ldots, 0, 1, 0, \ldots, 0, -1, 0, \ldots, 0) \end{aligned}$$

(the 1 in the k-th entry and the $-1$ in the j-th). We observe that for $s \le k$, the sum of the first $s$ entries of $a^h - a$ is not positive. Also, $\sum_{s\le k} a_s = \sum_{s\le d} a_s$, but the $j = h(i)$-th entry of $a^g$ is $a_o$, so that (as $k < h(i)$), $\sum_{s\le k}(a^h - a)_s \le -a_i$; by the same reasoning, for any $t < h(i)$, $\sum_{s\le t}(a^h - a)_s \le -a_i$. Set $c = a^g - a - a_i e^d$. Then $c = (a^h - a) + a_i(0, \ldots, 0, 1, 0, \ldots, -1, 0, \ldots, 0)$ (1 in the k-th position, $-1$ in the h(i)-th), so for $t < k$, $\sum_{s\le t} c_s \le 0$; as $\sum_{s\le t}(a^h - a)_s \le -a_i$ for $t < h(i)$, we deduce $\sum_{s\le t} c_s \le 0$ for $t < h(i)$. Finally, $\sum_{s\le t} c_s = \sum_{s\le t}(a^h - a)_s$ for $t \ge h(i)$.

Thus $\sum_{s\le t} c_s \le 0$ for $t = 1, 2, \ldots, d-1, d$, and $\sum_{s\le d} c_s = 0$. Moreover as $(a^h - a)_s \le 0$ for $s > k$, we have that $c_s \le 0$ for $s > k$. Thus $c$ satisfies all of the properties attributed to $(b_1, \ldots, b_d)$ in the first paragraph of the proof of (b), so that $c$ is a positive integer combination of $\{e^1, \ldots, e^{d-1}\}$. As $a^g - a = c + a_i e^d$, $a^g - a$ is thus also a positive combination, this time allowing $e^d$ to appear.

Now we compute the volume of the polytope $V = \mathrm{cvx}\{\mathbf{0}, e^1, \ldots, e^d\}$. The determinant of the matrix whose rows are respectively $e^1, \ldots, e^{d-1}$, is exactly $(-1)^d(d+1)$; this may be seen by using the obvious column operations (add the first row to the second, the second to the third, etc.). Hence $V$ has volume $(d+1)/d!$; as $L = L_a - a$ is of index $d+1$ in $\mathbf{Z}^d$, it easily follows that $\{e^1, \ldots, e^d\}$ is a $\mathbf{Z}$-basis for $L$. Hence $V$ is actually the peak polytope for $K_a - a$ at $\mathbf{0}$ with respect to $L$, and with respect to this basis for $L$, $V$ is the standard simplex. As the other vertices of $K_a$ are obtainable from $a$ via a permutation (which is affine), it follows that $K_a$ is integrally simple with respect to $L$.

To deal with case (iii), we observe that the adjoint induces an involution on the representation ring of $G$ via $\pi \mapsto \pi^*$, where $\pi^*(g)$ is the complex conjugate of $\pi(g^{-1})$, for a representation $\pi$

of G. This has the effect of sending the irreducible character having dominant weight $a = (a_1,...,a_d)$ to the irreducible with dominant weight $\rho(a) = (a_1, a_1-a_d, a_1-a_{d-1},..., a_1-a_2)$. This extends to $\rho: \mathbf{Z}^d \longrightarrow \mathbf{Z}^d$, which normalizes the action of the Weyl group, and induces an integral affine homeomorphism $K_a \longrightarrow K_{\rho(a)}$. It is easy to see that the dominant weight $a$ satisfies condition (ii) if and only if $\rho(a)$ satisfies (iii), and vice versa. Thus condition (iii) is sufficient for $K_a$ to be L-integrally simple.

We say that the dominant weight $a$ is properly ordered if it satisfies (i) or (ii), and it is $\rho$-properly ordered if $a$ satisfies (iii). We must show that if $a$ is neither properly ordered, nor $\rho$-properly ordered, then $K_a$ is not simple as a real polytope. The proof goes by induction on $d$. First suppose that $a$ has an internal repetition, i.e., there exists $k$ with $d-2 \geq k \geq 1$ so that $a_k > a_{k+1} = a_{k+2} > a_d$.

Let $u$ denote the vector $u = (1,..., 1)$, and define the $d-1$-face $F_a = \{K_a \cap \mathbf{R}^d | u \cdot v = \sum a_i\}$. Consider the restriction of $S_{d+1}$ to this face, that is, consider elements of $S_{d+1}$ that leave $F$ stable; this corresponds to an action of $S_d$ on $\mathbf{Z}^{d-1}$ (obtained as a quotient of the permutation action of $S_d$ on $\mathbf{Z}^d$) and and after translating to the standard copy of $\mathbf{Z}^{d-1}$, the original $a$ is transformed into $a' = (a_1-a_d, a_2-a_d, ..., a_{d-1}-a_d)$, corresponding to a representation of SU(d), and to $K_{a'}$. In any case, we may think of $F_a$ as corresponding to $K_{a'}$, and $S_d$ is acting on it.

If $a'$ itself has an internal repetition, then by induction on the dimension, $d$, $K_{a'}$ is not simple as a real polytope, so that $K_a$ would have a non-simple face, and therefore could not be simple itself. The only way $a'$ could fail to have an internal repetition would be if there exist $m \geq 1$ and $k < d-2$, so that $a_1 = a_2 = ...= a_m > ... > a_k = ... = a_{d-1} > a_d > 0$. However, in this case, $a'_{d-1} \neq 0$, so that neither is $a$ properly ordered nor $\rho$-properly ordered; by induction, $K_{a'}$ cannot be simple.

Thus in order that $K_a$ be simple, to is necessary that $a$ admit no internal repetitions. Suppose that this is the cae, that is, only external repetitions occur (meaning $a_i = a_j$ for $i < j$ implies that either $a_i = a_1$ or $a_j = a_d$). If $a_1 > a_2$ (no initial repetition), then $a_1 > a_2 > ...> a_k = ... > a_k = ... = a_{d-1} = a_d$, with $k+1 < d$; when $a_d = 0$, $a$ is already properly ordered. We must show that if $a_d$ is not zero, then $K_a$ is not simple. We mimic the construction of the $e^i$ earlier in the proof.

Define

$$e^1 = (-1, 1, 0, \ldots, 0, 0, 0) \qquad f^{k+1} = (-1,-1,\ldots,-1,-2,-1,\ldots,-1,\ldots,-1).$$

$$e^2 = (0, -1, 1, 0, \ldots, 0, 0) \qquad f^{k+2} = (-1,-1,\ldots,-1,-1,-2,\ldots,-1,\ldots,-1).$$

$$\vdots \qquad\qquad \vdots$$

$$e^k = (0,0,\ldots,0,-1,1,0,\ldots,0,0) \qquad f^d = (-1,-1,\ldots,-1,-1,\ldots,-1,-2)$$

$$e^{k+1} = (0,0,\ldots,0,-1,0,1,\ldots,0,0)$$

$$\vdots$$

$$e^{d-1} = (0,0,\ldots,0,-1,0,\ldots,0,1)$$

$$e^d = (-1,-1,\ldots,-1,-2,-1,\ldots,-1,\ldots,-1).$$

Set $f^k = e^d$ (so that $f^{k+1}$ has its $-2$ one position to the right of that of $e^d$); this is also $f^{k+1} + e^k$. First, each of the e's and f's lies in $(K_a-a) \cap (L_a-a)$; this is proved as in the earlier argument, only now that $a_{k+1}$ through $a_d$ are not zero, we also acquire the f's (previously, only the e's were obtained). Next, we show that everything of the form $a^g-a$ ($g$ in $S_{d+1}$) is a non-negative integer combination of the elements of the set $\{e^i, f^j\}$. As before, there is a permutation h on d symbols so that $a^g-a = a^h-a - a_i e_{h(i)} - a_i(-1,-1,\ldots,-1)$ for some i. If $h(i) \geq k+1$, then $a^g-a = a^h-a + a_i f^{h(i)}$, and as a is a non-negative integer combination of the e's (as in the earlier argument), we are done in this case. If $h(i) = k$, then $a^g-a = a^h-a + a_i e^d$; if $j = h(i) < k$, then $a^g-a = a^h-a + a_i(e^d + e^j + \ldots + e^{k-1})$ and we are done.

This means that any edge containing $\mathbf{0}$ must be one of the line segments joining $\mathbf{0}$ to an $e^i$ ($i \neq d$) or an $f_j$ ($j < d$). We already know that $e^1, \ldots, e^{d-1}$ lie along such edges, because these affinely span the face $F_a-a$ (defined by the relation, the sum of the coefficients is zero). Next, we show that none of the f's (including $f^k = e^d$) can be deleted without reducing the real cone that they span. If some $f^j$ ($j > k$) could be deleted, we could write with the λ's and μ's in $\mathbf{R}^+$:

$$f^j = \sum_{i \leq d-1} \lambda_i e^i + \sum_{k < s \neq j} \mu_s f^s \qquad (1)$$

Taking the sum of the coefficients of the entries, we see that the sum of the relevant $\mu_s$ terms is 1. Hence for $t<k$, the t-th coefficient of $\sum_{s \neq j} \mu_s f_s$ is $-1$; as $(f^j)_t = -1$ as well, we deduce

$(\sum_{i<d} \lambda_i e^i)_t = 0$ for $t < k$. This entails $\lambda_1 = 0$, which yields $\lambda_2 = 0$, and so on until we obtain $\lambda_{k-1} = 0$. Thus

$$(f^j)_k = -\sum_{d>i\geq k} \lambda_i + \sum \mu_s \cdot (f^s)_k .$$

However, $(f^j)_k = (f^s)_k = -1$ as j and s exceed k. Since $\sum \mu_s = 1$, we have that

$$-1 = \sum_{d>i\geq k} \lambda_i + (-1)$$

so that $\sum \lambda_i = 0$, and thus all the remaining $\lambda$'s are zero as well. Thus (1) degenerates to:

$$f^j = \sum_{k<s\neq j} \mu_s f^s. \qquad (1')$$

Subtracting $e_{d+1} = (-1, ..., -1)$ from both sides yields:

$$f^j - e_{d+1} = \sum \mu_s \cdot (f^s - e_{d+1}). \qquad (1'')$$

However, the left hand side is $(0,..., 0,-1, 0,..., 0) = -e_s$, so there results a violation of the linear independence of $\{e_s\}_{d\geq s>k}$.

Hence each of the f's correspond to an edge of $K_a - a$ that contains $\mathbf{0}$. Thus there are at least (in fact the argument yields exactly) $d-k+d-1 > d$ such edges, so that $K_a - a$ is not simple.

The only remaining case concerns the possibility that a admit no internal repetitions, but $a_1 = a_2$. If there is also a final repetition, i.e., $a_1 > a_{d-1} = a_d$, then either $a_d$ is not zero or it is zero. If it is zero, then $\rho(a)_1 > \rho(a)_2 = \rho(a)_3 \neq 0$ and thus $\rho(a)$ either has an internal repetition or otherwise violates conditions already known to result in non-simplicity. If $a_1 = a_2 > a_{d-1} = a_d > 0$, then $a' = (a_1, ..., a_{d-1} = 0)$ and a' has an initial repetition, so by induction, $K_{a'}$ is not simple, and thus neither is $K_a$.

Finally, if $a_1 = a_2$ and $a_{d-1} > a_d > 0$, and a admits no internal repetitions, then a is $\rho$-properly ordered, so that all of the cases have now been considered.∎

*A few dihedral groups*

Let $D_n$ denote the dihedral group of order $2n$. We shall show that for $n = 6$ or 4, there are up to conjugation by elements of $GL(2,\mathbf{Z})$, exactly two faithful actions of $D_n$ on $\mathbf{Z}^2$ and these are outer conjugate, i.e., there exists an outer automorphism of the group sending one (equivalence class) of) representation(s) into the other. In particular, there is only one set of orbits to look at, and we show that for these representations, integral simplicity of $K_a$ (possibly with respect to a sublattice of $\mathbf{Z}^2$) holds for almost all choices of a.

First let us consider the faithful representations of $D_6$ into $GL(2,\mathbf{Z})$. The element of order 6, call it g, must be sent to a $2 \times 2$ integral matrix whose characteristic polynomial is $x^2 - x + 1$, so it has eigenvalues $-w$ and $-w^2$ where w is the primitive cube root of unity, $w = \exp(2\pi i/3)$. As the ring $\mathbf{Z}[-w] = \mathbf{Z}[w]$ is integrally closed and a principal ideal domain, there is only one conjugacy class (with respect to $GL(2,\mathbf{Z})$) of such matrices. So we may assume that g is sent by the representation to $\left[\begin{smallmatrix} 1 & 1 \\ -1 & 0 \end{smallmatrix}\right]$ in $GL(2,\mathbf{Z})$.

Now let h denote the order two element of $D_6$ with the property that $hg = g^{-1}h$. Suppose h is sent to the matrix $\left[\begin{smallmatrix} a & b \\ c & d \end{smallmatrix}\right]$ in $GL(2,\mathbf{Z})$. As $h^2 = 1$, we deduce $d = -a$ and $a^2 + bc = 1$. From the relation $hg = g^{-1}h$, we deduce further that $c = b - a$. Substituting for c into $a^2 + bc = 1$, we deduce $a^2 + b^2 - ab = 1$, or in other words $(a - b/2)^2 + 3b^2/4 = 1$. As a and b are integers, we deduce that the only possible values for b are $0, \pm 1$. Each value of b yields exactly two values for a, so we obtain 6 possible matrices that h may be sent to. However, we may conjugate each of these with the image of g without affecting the relation, $hg = g^{-1}h$. We quickly see that that there are precisely two orbits in the 6 matrices under this action; representatives of these orbits are determined by $h \mapsto \left[\begin{smallmatrix} 0 & 1 \\ 1 & 0 \end{smallmatrix}\right]$ and $h \mapsto \left[\begin{smallmatrix} 0 & -1 \\ -1 & 0 \end{smallmatrix}\right]$. Since $\mathbf{Z}[w]$ is integrally closed, and its only units $1, w, -w, w^2, -w^2$, and $-1$, it easily follows that the corresponding representations of $D_6$ are inequivalent (for example, the centralizer in $GL(2,\mathbf{Z})$ of the image of g consists of the polynomials in g with integer coefficients).

Next we notice that these two representations give rise (in their actions on $\mathbf{Z}^2$) to identical orbits, because there is an automorphism of $D_6$ that fixes g and sends h to $g^3h$; this interchanges the two representations. Since we are only interested in the orbits, we may as well assume that we are working with the representation given by :

$$g \mapsto \begin{bmatrix} 1 & 1 \\ -1 & 0 \end{bmatrix}, \quad h \mapsto \begin{bmatrix} 0 & 1 \\ 1 & 0 \end{bmatrix}.$$

Under these circumstances, we can easily compute the orbit of a given point $(a,b)^T$ (superscript T for *transpose*) in $\mathbf{Z}^2$. Its orbit under powers of g is given by the transposes of the members of the set:

$$\{(a,b),\ (a+b,-a),\ (b,-a-b),\ (-a,-b),\ (-a-b,a),\ (-b,a+b)\};$$

that of h just flips the two entries (so (a,b) is replaced by (b,a), and so on). We may assume that $a \geq b \geq 0$. If $a \neq b$ and $b \neq 0$, then the two points nearest the vertex (a,b) of the convex hull of the orbit are (b,a) and (a+b,–b). Subtracting (a,b) from these three points, we arrive at the situation in diagram D1:

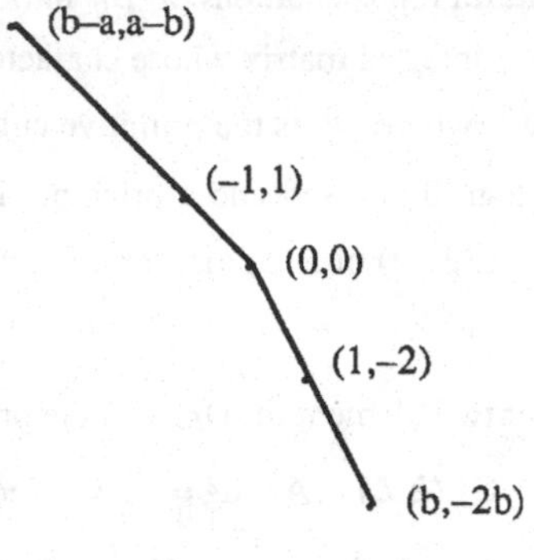


Diagram D1

We observe that on the line joining (b–a,a–b) to the origin is the point (–1,1); similarly on the other line segment is the point (1,–2). The area of the triangle with vertices {(0,0), (–1,1), (1,–2)} is $\frac{1}{2}\det\begin{bmatrix} -1 & 1 \\ 1 & -2 \end{bmatrix} = \frac{1}{2}$. Thus if we let $L = \mathbf{Z}^2$ (the standard lattice), then $K_a$, the convex hull of the orbit of (a,b), is integrally simple.

If $a = b \neq 0$, then the orbit is only a hexagon (as opposed to a duodecagon as in the previous case), and the corresponding diagram has vertices {(0,0), (–2a,–a), (a,–2a)}, which reduce to {(0,0), (–2,–1), (1,–2)}. Now $\det\begin{bmatrix} -2 & 1 \\ 1 & -2 \end{bmatrix} = 3$, so the hexagon is not integrally simple.

Finally, if $a \neq b = 0$, then the corresponding vertices are {(0,0), (–a,a), (0,–a)}; reduced, this becomes the set of vertices, {(0,0), (–1,1), (0,–1)} and the corresponding triangle obviously has area $\frac{1}{2}$, so here too $K_a$ is integrally simple.∎

The case of $D_4$ is so similar, that we merely outline what to do. First we note that $\mathbf{Z}[i]$ is principal, so if $D_4$ is generated by g and h where $g^4 = 1$, $h^2 = 1$, and $gh = hg^{-1}$, then the image of g is unique up to conjugation by GL(2,Z), so we may assume g is sent to $\begin{bmatrix} 0 & 1 \\ -1 & 0 \end{bmatrix}$. Then use the equations to deduce (in analogy with the previous example) that h must be sent to one of

$$\begin{bmatrix} 0 & 1 \\ 1 & 0 \end{bmatrix}, \begin{bmatrix} 0 & -1 \\ -1 & 0 \end{bmatrix}, \begin{bmatrix} 0 & 1 \\ 1 & 0 \end{bmatrix}, \begin{bmatrix} -1 & 0 \\ 0 & 1 \end{bmatrix}.$$

Conjugation by the image of g yields two 2-element orbits, and as above, the corresponding two representations are inequivalent, but outer conjugate. Then we just compute the 8 (or occasionally 4)-element orbits of the action given by the group generated by $\begin{bmatrix} 0 & 1 \\ -1 & 0 \end{bmatrix}$ and $\begin{bmatrix} -1 & 0 \\ 0 & 1 \end{bmatrix}$ on $\mathbf{Z}^2$, computing determinants as above, and arrive at all the convex hulls of the orbits being integrally simple (i.e., with respect to $\mathbf{Z}^2$) . The nearest neighbours to (a,b) among the points in the orbit of (a,b), are (b,a) and (a,–b), except when either a = b or b = 0 (in the latter two situations, a square results).∎

If instead of considering the action of $D_6$ described above, we merely consider the action of the cyclic subgroup of order 6, then the corresponding convex hulls are rarely integrally simple. Thus the presence of many reflections seems to have something to do with integral simplicity.

We also remark that in the cases of $D_6$ and $D_4$, it was natural to take as the lattice with respect to which we were checking for integral simplicity, the whole of $\mathbf{Z}^2$, in contrast to the cases considered in Theorems D1 and D2. For other reflection groups, there are a number of choices for the appropriate lattice; some will work and others will not.

To illustrate this, we consider $D_3$. There is one integral representation (on $\mathbf{Z}^2$) arising from it being the Weyl subgroup of SU(3) (this is the one occurring in D1 for SU(3)). This action leads to a sublattice $L_o$ of index 3 in $\mathbf{Z}^2$. There is up to GL(2,Z)-conjugacy exactly one other faithful representation on $\mathbf{Z}^2$, given by

$$g \mapsto \begin{bmatrix} -1 & 1 \\ -1 & 0 \end{bmatrix}, \qquad h \mapsto \begin{bmatrix} 0 & -1 \\ -1 & 0 \end{bmatrix}.$$

It is routine to verify (as in the previous computations) that $K_a$ is integrally simple with respect to $\mathbf{Z}^2$ itself for all a in $\mathbf{Z}^2$. In particular, this geometric property allows one to distinguish equivalence classes of representations.

## Appendix E. THE LEGENDRE TRANSFORMATION

Here we prove, without convex analysis, that the map $\Lambda:(\mathbf{R}^d)^{++} \longrightarrow \text{Int } K$ described in section IV is a homeomorphism. In fact, we prove a more general result, but do not obtain the full generality of [Ro; Theorem 26.5]. Recall that if $P = \sum \lambda_w x^w$ and $K = \text{cvx Log } P$, $\Lambda$ is defined via

$$\Lambda(r) = \sum_{w \in \text{Log } P} \frac{\lambda_w r^w w}{P(r)}.$$

It turns out that this is a special case of a more general phenomenon. Let $C$ be a compact convex subset of $\mathbf{R}^d$ having interior, and let $\mu$ be a (Borel) probability measure on $C$ such that the closed convex hull of the support of $\mu$ is $C$ itself. Using the exponential map $\mathbf{R}^d \longrightarrow (\mathbf{R}^d)^{++}$, we define a map analogous to $\Lambda$, $\Gamma_\mu:\mathbf{R}^d \longrightarrow \text{Int } C$ given by

$$s \mapsto \frac{\int_C w \exp(s\cdot w)\, d\mu(w)}{\int_C \exp(s\cdot v)\, d\mu(v)}.$$

The expression $s\cdot w$ is of course the inner product (we sometimes suppress the dot). If $\mu$ is atomic with finite support $S \subset \mathbf{Z}^d$ and say $\mu(\{w\}) = \lambda_w$ for each $w$ in $S$, then after exponentiating $\mathbf{R}^d$, we recover our original map $\Lambda$.

Even in this generality, we shall prove:

(*) If $C$ contains interior, then $\Gamma_\mu$ is a homeomorphism onto Int $C$.

$\Gamma_\mu$ ***is one to one and open:***

Here we prove that when the hypothesis of (*) holds, $\Gamma_\mu$ is one to one and open. Owing to some computations involving first derivatives, it is easier to deal with the situation arising when $\Gamma_\mu$ is replaced by $\Lambda_\mu:(\mathbf{R}^d)^{++} \longrightarrow C$, where

$$\Lambda_\mu(r) = \frac{\int_C w\, r^w\, d\mu(w)}{\int_C r^w\, d\mu(w)}.$$

We remark that if $E:\mathbf{R}^d \longrightarrow (\mathbf{R}^d)^{++}$ is the exponential map given by $E(s) = \exp(s)$, then $\Gamma_\mu = \Lambda_\mu \circ E$.

Shifts act naturally upon the maps $\Gamma_\mu$ and $\Lambda_\mu$ in the following sense:

(Translation principal)E1 **LEMMA** Let $C$ be a compact convex subset (with interior) of $\mathbf{R}^d$, $v$ an element of $\mathbf{R}^d$, and $\mu$ a measure on $C$. Let $\mu'$ be the corresponding measure on $C + v$, that is, $\mu'(A) = \mu(A-v)$ for $A$ a Borel subset of $C + v$. Then the two maps $\Lambda_\mu:(\mathbf{R}^d)^{++} \longrightarrow C$ and $\Lambda_{\mu'}:(\mathbf{R}^d)^{++} \longrightarrow C + v$ are related via $\Lambda_\mu(r) + v = \Lambda_{\mu'}(r)$.

Proof: Indeed,
$$\Lambda_{\mu'}(r) = \frac{\int_{C+v} u r^u \, d\mu'(r)}{\int_{C+v} r^u \, d\mu'(u)}$$

$$= \frac{\int_C (w+v) r^w r^v \, d\mu(w)}{\int_C r^w r^v \, d\mu(w)}.$$

The terms in $r^v$ cancel, and we obtain

$$\Lambda_{\mu'}(r) = \Lambda_\mu(r) + v\frac{\int_C r^w \, d\mu(w)}{\int_C r^w \, d\mu(w)} = \Lambda_\mu(r) + v. \blacksquare$$

Of course a similar property holds for $\Gamma_\mu$. As a consequence, we can often perform a translation (on $C$) without affecting properties of $\Lambda_\mu$ or $\Gamma_\mu$. For example (as will be seen below in the proof that the maps are one to one), we may assume that given $u$ in $\mathbf{R}^d \setminus \{\mathbf{0}\}$, $u \cdot c > 0$ for all $c$ in $C$. In the proof of ontoness, it is sufficient to show that $\Gamma_\mu(r) = \mathbf{0}$ can be solved whenever $\mathbf{0}$ is in the interior of $C$. We shall use this principal without much further comment.

More generally, if $g$ is an element of $AGL(d,\mathbf{Z})$ and $g(C) = C'$, then $g\Lambda_\mu = \Lambda_{g\mu}$ where $g\mu$ is the measure on $C'$ obtained via $(g\mu)(A) = \mu(g^{-1}A)$ when $A$ is a subset of $C'$. However, we only require that this type of property hold for translations.

Define a formal polynomial to be a function (of one variable) of the form:

$$p(t) = \sum_{i=0}^{N} a_i t^{\alpha(i)} \qquad (1)$$

with complex numbers $a_i$, where the $\alpha(i)$ are real numbers satisfying $0 \le \alpha(0) < \alpha(1) < \ldots < \alpha(N)$. Obviously $p$ is analytic in $\mathbf{C} \setminus \mathbf{R}^-$ (taking the principal value of $\exp(\alpha(i)\ln t)$ for $t^{\alpha(i)}$). If in expression (1), $a_N \neq 0$, we say $p$ has (formal) degree $\alpha(N)$. If all the coefficients, $a_i$ are real and positive, then we say $p$ has positive coefficients. The following is directly from [BW; p. 39] (although there is a minor typographical error there, and they consider only polynomials in the usual sense):

**E2 LEMMA** ([BW; p. 39]) If $p$ is a formal polynomial of formal degree $D = \alpha(N)$ having positive coefficients, and $(\ln p)'$ is not constant, then

$$t(tp''p + p'p - t(p')^2) = tp^2(tp'/p)'$$

is a formal polynomial of degree less than 2D, having positive coefficients.

Proof: Using (1), we see that for any real number (possibly negative) $\beta$, the coefficient of $t^\beta$ in $tp''p+p'p-t(p')^2$ is

$$\sum_{(i,j)\in S(\beta)} a_i a_j \alpha(i)(\alpha(i)-\alpha(j)),$$

where $S(\beta) = \{(i,j) | \alpha(i) + \alpha(j) = \beta + 1\}$. Interchanging $i$ and $j$ in this sum, we obtain that the coefficient is $\frac{1}{2}\sum_{S(\beta)} a_i a_j (\alpha(i) - \alpha(j))^2$. Hence the coefficients are all non-negative. The least $\beta$ that can occur (with nonzero coefficient is $-1$, and this only if $\alpha(0) = 0$; in fact, the least such $\beta$ must exceed $-1$, as it corresponds to $i = j = 0$. Similarly the largest exponent that can appear is $\alpha(N) + \alpha(N-1) - 1$, and this actually does appear. Multiplying by $t$ will render it a formal polynomial of degree $\alpha(N) + \alpha(N-1) < 2\alpha(N)$.∎

**E3 LEMMA** If $Q$ is a formal polynomial of degree $D = \alpha(N)$ having positive coefficients, and if the (non-real complex number $z_o = \rho_o \exp(i\theta_o)$ (with $-\pi < \theta_o < \pi$) is a root of $Q$, then $|D\theta_o| > \pi$.

Proof: As the imaginary part of $Q(z_o)$ is given by

$$\operatorname{Im} Q(z_o) = \sum a_j \rho_o^{\alpha(j)} \sin(\alpha(j)\theta_o),$$

and the coefficient of each $\sin(\alpha(j)\theta_o)$ is positive, the set $\{\sin(\alpha(j)\theta_o)\}$ must contain some sign changes. Obviously this can happen only if $|D\theta_o| > \pi$.∎

**E4 LEMMA** Let $\{Q_n\}$ be a sequence of formal polynomials with positive coefficients and of bounded formal degrees, converging uniformly on a compact neighbourhood $V$ (in $\mathbf{C}$) of the positive real number $r_0$, to a function $Q$ (which is necessarily analytic on Int V). Then there exists a compact neighbourhood $W \subset V$ of $r_0$ such that no $Q_n$ has a zero in W. In particular, either Q is identically zero on V, or $Q(z) \neq 0$ for all z in Int W.

Proof: If R is a positive real number exceeding the degrees of all of the $Q_n$'s, set $W = V \cap \{z \in \mathbf{C} \mid |\mathrm{Arg}\, z| < \pi/R\}$. By the previous result, no $Q_n$ can vanish on W. The final statement is in [A; p. 176].∎

**E5 LEMMA** Let $\{P_n\}$ be a sequence of formal polynomials with positive coefficients and of bounded formal degrees, that converges uniformly on a compact neighbourhood (in **C**) of the positive real number $r_0$ to the function P. If P and Log P are not constant or linear, then there exists a neighbourhood U of $r_0$ (in **C**) such that $(tP'/P)'$ does not vanish at any point of U.

Proof: Of course P is analytic on an open neighbourhood of $r_0$, and $\{P'_n\}$ converges uniformly to $P'$ on the original compact neighbourhood of $r_0$ [A; p. 174]. With $P_n = Q_n$ and $Q = P$, by E4, we may find a compact subneighbourhood W such that P does not vanish on it. Hence $\{tP'_n/P_n\}$ converges uniformly to $tP'/P$ on W, and thus the derivatives do as well. By E3, $tP_n^2(tP'_n/P_n)'$ are formal polynomials of bounded degree with positive coefficients, and they obviously converge to $tP^2(tP'/P)'$. Setting $Q_n = tP_n^2(tP'_n/P_n)'$, and applying the previous lemma completes the proof.∎

Let $C,\mu$ be a compact convex subset of $\mathbf{R}^d$ with a probability measure, such that the closed convex hull of the support of $\mu$ is all of C. Also assume that C contains an open d-ball (so C is a "convex body"). Define a function $P:(\mathbf{R}^d)^{++} \longrightarrow \mathbf{R}^+$ via

$$P(r) = \int_C r^w \, d\mu(w).$$

Then P extends to a function with domain $\{z \in \mathbf{C}^d \mid \mathrm{Re}\, z >> (0,...,0)\} = U$, and range in **C** by means of the same formula. This extension is obviously analytic since any approximation of $\mu$ by atomic probability measures on C will lead to an approximation of P, uniform on compact neighbourhoods in U, by formal polynomials in several variables. In particular, the map $\Lambda_\mu$ extends to an analytic function from U to $\mathbf{C}^d$ with the formula:

$$\Lambda_\mu(z) = \int_C w(z^w/P(z))\, d\mu(w).$$

As $r^w/P(r)$ strictly exceeds 0 (for r in $(\mathbf{R}^d)^{++}$), it easily folows that $\Lambda_\mu((\mathbf{R}^d)^{++})$ is contained in the interior of C. We can finally prove that $\Lambda_\mu$ is one to one (on restriction to $(\mathbf{R}^d)^{++}$):

**E6 PROPOSITION** The function $\Lambda_\mu:(\mathbf{R}^d)^{++} \longrightarrow \mathrm{Int}\, C$ is one to one.

Proof: Select two distinct points $r = (r_1, r_2, ..., r_d)$ and $r' = (r'_1, r'_2, ..., r'_d)$ in $(\mathbf{R}^d)^{++}$; we wish to show that $\Lambda_\mu(r) \neq \Lambda_\mu(r')$. Set $A_i = r_i$, $u(i) = \ln(r'_i/r_i)$, and define the path $X:(0,\infty) \longrightarrow (\mathbf{R}^d)^{++}$,

$$X(t) = (..., A_i t^{u(i)}, ...).$$

Note that $X(1) = r$ and $X(e) = r'$, and that $X(t)$ is the exponential of the straight line $\log r + su$ ($s$ in $\mathbf{R}$) where $u = (u(1), ..., u(d))$.

Define $f(t)$ to be $u\cdot((\Lambda_\mu \circ X)(t))$. This defines $f$ as a real analytic function from $(0,\infty)$ to $\mathbf{R}$. Explicitly, on setting

$$Q(t) = \int_C t^{u\cdot w} A^w d\mu(w),$$

we have that

$$f(t) = \frac{\int_C (u\cdot w) t^{u\cdot w} A^w d\mu(w)}{Q(t)} = \frac{t Q'(t)}{Q(t)}.$$

We wish to show that $f'(t) > 0$ for all $t > 0$. We first observe that we may have added a vector to $C$ and so translated $C$ so that $u\cdot c > 0$ for all $c$ in $C$, without affecting (for example) the non-vanishing of the derivative. We may thus assume that all the $u\cdot w$'s appearing above are non-negative, and since $C$ contains interior, $u\cdot w$ takes on at least an interval of values as $w$ varies over $C$.

As $\mu$ is approximated by atomic measures, $Q(z)$ will correspondingly be approximated uniformly on compact subsets of $\{z \in \mathbf{C} | \operatorname{Re} z > 0\}$ by formal polynomials, and as the values of $u\cdot w$ are non-negative and bounded on $C$, we may take these formal polynomials to have positive coefficients, and have their degrees bounded.

By E5, $f'(t) > 0$ for all $t > 0$. However, $\Lambda_\mu(r) = \Lambda_\mu(r')$ would entail $f(1) = f(e)$, which is obviously impossible. Hence $\Lambda_\mu$ is one to one.∎

**E7 PROPOSITION** The function $\Lambda_\mu$ is open and a local diffeomorphism.

Proof: It is clearly equivalent to show that the desired properties hold for $\Gamma_\mu$. Let $J(s)$ denote the Jacobian of $\Gamma_\mu$ at $s$ in $\mathbf{R}^d$. To show that $J(r)$ is not zero, it is sufficient to show that given arbitrary elements $a$ and $u$ of $\mathbf{R}^d$, there is a differentiable path $Y: \mathbf{R} \longrightarrow \mathbf{R}^d$ with $Y(0) = a$, $Y'(0) = u$, and such that $\frac{d}{ds}(\Gamma_\mu \circ Y)(t)$ does not vanish at any point $t$ in $\mathbf{R}$ (I am indebted to Wulf Rossmann for this remark). Setting $Y(t) = a + ut$, we see that if $X(t')$ is defined (for $t' = \exp(t)$

and with $A_i = \exp(a(i)))$ as $\exp Y(t) = (\ldots, A_i t^{u(i)}, \ldots)$ then $X$ is just the type of path discussed in the argument above. We thereby have $u \cdot ((\Lambda_\mu \circ X)(t)) \neq 0$, and a simple application of the chain rule yields non-vanishing of $\frac{d}{ds}(\Gamma_\mu \circ Y)(t)$. Hence $\Gamma_\mu$ open and locally a diffeomorphism.∎

$\Gamma_\mu$ ***is onto:***

As was remarked earlier, the range of $\Lambda_\mu$ (and $\Gamma_\mu$) is in the interior of C; this also follows from the map being open. In this section, it is more convenient to work directly with $\Gamma_\mu$.

E8 **LEMMA** Let $\{r^k\}$ be an unbounded sequence in $\mathbf{R}^d$. Then all limit points of $\{\Gamma_\mu(r^k)\}$ in C that cannot be obtained as limits of subsequences arising from a bounded subsequence of $\{r^k\}$, must lie in boundary of C.

<u>Proof</u>: Suppose not; we may refine the sequence $\{r^k\}$ and so assume that $\lim_{k\to\infty} \Gamma_\mu(r^k)$ exists and equals c in the interior of C. Define probability measures $\mu_k$ on C (with A a Borel subset of C):

$$\mu_k(A) = \frac{\int_A \exp(r^k \cdot w)\, d\mu(w)}{\int_C \exp(r^k \cdot v)\, d\mu(v)} .$$

Then $\Gamma_\mu(r^k) = \int_C w\, d\mu_k(w)$. Consider the limit points of $\{\mu_k\}$ in the weak topology (on measures, i.e., on positive linear functionals of norm 1 of $C(C,\mathbf{R})$ ). There is a limit point $\nu$ such that $c = \int_C w\, d\nu(w)$. We will show that the support of $\nu$ is contained in a convex subset of the boundary (actually its convex hull is a proper face of C), and this last equation then obviously yields that c belongs to the boundary of C, a contradiction.

Form a sequence of bounded elements in $\mathbf{R}^d$, $\{u^k = r^k/\|r^k\|_2\}$; this has a limit point u which is a unit vector in $\mathbf{R}^d$. By taking a suitable subsequence of $\{r^k\}$, we may assume that $\lim_{k\to\infty} u^k = u$ (in addition to $\lim \Gamma_\mu(r^k) = c$). Set $a = \sup\{u \cdot w \mid w \in C\}$ and $F = \{c \in C \mid u \cdot c = a\}$ (so F is a face of C). Let b be a real number less than a, and set $A = \{w \in C \mid u \cdot w \leq b\}$. Select a real number e so that $a > e > b$, and define B as $\{w \in C \mid e \leq u \cdot w \leq a\}$. We shall show that $\nu(A) = 0$.

By a translation (if necessary) of C, we may assume that $a > e > 0 > b$ (note that since C contains interior, u cannot be constant on C, i.e., $F \neq C$). We have:

$$\mu_k(A) = \frac{\int_A \exp(r^k \cdot w)\, d\mu(w)}{\int_C \exp(r^k \cdot w)\, d\mu(w)}$$

$$\leq \frac{\mu(A)}{\mu(B)} \cdot \frac{\sup\left\{\exp(r^k \cdot w) \mid w \in A\right\}}{\inf\left\{\exp(r^k \cdot w) \mid w \in B\right\}} \qquad (2)$$

Now choose $\varepsilon > 0$. There exists $k'$ so that for all $k$ exceeding $k'$, $\|u^k - u\|_2 < \varepsilon$. Let $M = \sup\{\|w\|_2 \mid w \in C\}$. Then $|u^k \cdot w - u \cdot w| < \varepsilon M$ for all $w$ in $C$ and all sufficiently large $k$. If we select $\varepsilon$ small enough so that $\varepsilon M < |b|$ (recall: $b < 0$), then we obtain $u^k \cdot w < u \cdot w + \varepsilon M$ for all $w$ in $A$. Hence $r^k \cdot w < 0$ for all $w$ in $A$. Thus the numerator of the expression in (2) is bounded above by $\nu(A)$, for all sufficiently large k.

We also have $u^k \cdot w > u \cdot w - \varepsilon M$. We may further reduce $\varepsilon$ and correspondingly alter $k'$ so that $\varepsilon M > e/2$. Thus for $w$ in B, $u^k \cdot w > b/2$. For such $w$, $r^k \cdot w > \frac{1}{2}\|r^k\|$. So the denominator of the expression in equation (2) is bounded below by $\mu(B)\exp(\frac{1}{2}\|r^k\|)$ for all sufficiently large $k$. As $B$ is a compact neighbourhood of $F$, and the latter is a face in $C$, our hypothesis that the closed convex hull of the support of $\mu$ is $C$ yields that $\mu(B) \neq 0$. Thus as $k$ tends to infinity, the numerator remains bounded, while the denominator becomes arbitrarily large. Thus $\mu_k(A)$ tends to $0$ as $k$ increases, so that if $\nu$ is any limit point of $\{\mu_k\}$, we deduce that $\nu(A) = 0$. Now translate $C$ back to its original position.

If we allow $b$ to run over the rationals approximating $a$ from below and obtain sets $A_b$ corresponding to $A$, we see that each $\nu(A_b)$ is 0. As $\cup A_b$ is $C \setminus F$, we have that $\nu(F) = 1$, as desired.∎

The outcome of this argument is that if $\{r^k\}$ is an unbounded sequence in $\mathbf{R}^d$ with no bounded subsequences, then any limit point of $\{\Gamma_\mu(r^k)\}$ lies in the boundary of C. The portion of the conclusion of the argument that contains an estimate for $\int_C \exp(r^k \cdot w)\, d\mu_k(w)$ yields the following:

E9 **PROPOSITION** If $\mathbf{0}$ lies in Int C, and $\{r^k\}$ is an unbounded sequence in $\mathbf{R}^d$, then $\left\{\int_C \exp(r^k \cdot w)\, d\mu_k(w)\right\}_{k \to \infty}$ is unbounded.

E10 **PROPOSITION** If $\mathbf{0}$ lies in Int C, then $\mathbf{0}$ lies in the range of $\Gamma_\mu$. (By our translation principal, E1, this is sufficient to show that $\Gamma_\mu$ maps onto Int C.)

<u>Proof</u>: Define $N:\mathbf{R}^d \to \mathbf{R}^d$ via $N(x) = \int_C w \exp(x\cdot w)\,d\mu(w)$. It is sufficient to solve the equation $N(x) = \mathbf{0}$, as for all $r$ in $\mathbf{R}^d$, $N(r) = \Gamma_\mu(r)\int_C \exp(r\cdot w)\,d\mu(w)$. To this end, we define $F:\mathbf{R}^d \to \mathbf{R}^+$ as

$$F(x) = \|N(x)\|_2^2 = \sum_{i=1}^{d}\left(\int_C w(i)\exp(x\cdot w)\,d\mu(w)\right)^2.$$

We first show that $F$ attains its minimum in $\mathbf{R}^d$, and then that the value at the minimum is zero. Suppose $\{r^k\}$ is an unbounded sequence in $\mathbf{R}^d$ such that $\{F(r^k)\}$ is bounded. By E9, $\{\Gamma_\mu(r^k)\}$ converges to $\mathbf{0}$ (in the interior of C). By E8, this is impossible. Hence $\{F(r^k)\}$ unbounded whenever $\{r^k\}$ is. It follows easily that $F$ attains its minimum. Obviously $F$ is analytic.

Let $y$ in $\mathbf{R}^d$ be a point at which $F(y)$ is minimal. Then $\frac{\partial F}{\partial x_j}(y) = 0$ for $j = 1, 2, \ldots, d$. Define $F_i = \int_C w(i)\exp(y\cdot w)\,d\mu(w)$, so that for each value of $j$, the following holds:

$$0 = \tfrac{1}{2}\frac{\partial F}{\partial x_j}(y) = \sum_{i=1}^{d} F_i\left(\int_C w(i)\,w(j)\exp(y\cdot w)\,d\mu(w)\right) \qquad (3j)$$

We can write this system (linear in $F_i$) of equations as an operator equation. Define the following operators on the corresponding real Hilbert spaces:

| | | | | |
|---|---|---|---|---|
| $I_C:L^2(C) \to \mathbf{R}$ | via | $I_C(g)$ | = | $\int_C g(w)\,d\mu(w)$; |
| $\Delta: L^2C) \to L^2(C)$ | via | $\Delta(g)(w)$ | = | $g(w)\exp(y\cdot w)$; |
| $V: L^2(C) \to l^2\{1,2,\ldots,d\}$ | via | $V(g)(i)$ | = | $\int_C g(w)\,w(i)\,d\mu(w)$; |
| $\mathbf{F}: l^2\{1,2,\ldots,d\} \to \mathbf{R}$ | via | $\mathbf{F}(x)$ | = | $\sum x(i)F_i$; |

and for $j = 1, 2, \ldots, d$, define

| | | | | |
|---|---|---|---|---|
| $b_j:\mathbf{R} \to L^2(C)$ | via | $b_j(1)(w)$ | = | $w(j)$. |

These are obviously bounded linear operators with adjoints (indicated by *). We have the following operator equation:

$$\mathbf{F} = I_C\,\Delta\,V^*.$$

Moreover, the equations (3j) translate to:

$$\mathbf{F}\, V\Delta b_j = 0 \quad \text{for } j = 1,2,\ldots,d \qquad (3_j').$$

Now $V^*: l^2\{1,2,\ldots,d\} \longrightarrow L^2(C)$ is simply $(b_1, b_2, \ldots, b_d)$. Hence the equations $(3_j')$ yield:

$$\mathbf{F}\, V\Delta V^* = \mathbf{0}.$$

Next we observe that $\Delta = \Delta^*$ and this is a positive operator (its positive square root is multiplication by $\exp(\frac{1}{2} y\cdot w)$ ). Applying $\mathbf{F}^*$, we obtain $(\mathbf{F}\, V\Delta^{\frac{1}{2}})(\mathbf{F}\, V\Delta^{\frac{1}{2}})^* = \mathbf{0}$, whence

$$\mathbf{F}\, V\Delta^{\frac{1}{2}} = \mathbf{0}.$$

As $\Delta$ is invertible (its inverse is multiplication by $\exp(-y\cdot w)$ ), there results $\mathbf{F} V = \mathbf{0}$. Putting in our initial equation for $\mathbf{F}$ yields:

$$l_C \Delta V^* V = \mathbf{0}.$$

Hence $(l_C\Delta V^*)(l_C\Delta V^*)^* = \mathbf{0}$, and thus $l_C\Delta V^* = \mathbf{0}$, that is $\mathbf{F} = \mathbf{0}$. However, this simply says that each $F_i$ is zero.

Thus $F(r) = 0$ for some $r$ in $\mathbf{R}^d$, so that $\Gamma_\mu(r) = \mathbf{0}$, and the translation principal concludes the ontoness argument that $\Gamma_\mu$ (and thus $\Lambda_\mu$) is onto.∎

## REFERENCES


[A] L.V. Ahlfors, **Complex Analysis,** 2nd edition, McGraw-Hill, New York (1966).

[AI] R. Askey and M. Ismail, **Recurrence Relations, Continued Fractions and Orthogonal Polynomials,** Memoirs of the Amer. Math. Soc. **300** (1984) 107 p.

[Ba] H. Bass, **Algebraic K-Theory**, Benjamin (1968) New York.

[BH] B.M. Baker and D.E. Handelman, Time dependent integer-valued random walks, eventual positivity of polynomials, and the $K_0$-theory of product-type actions of the torus, preprint (1986).

[BW] W.G. Bardsley and R.M. Wood, Critical points and sigmoidicity of positive rational functions, Amer. Math. Monthly (1985) 37-49

[EHS] E.G. Effros, D.E. Handelman, and C.-L. Shen, Dimension groups and their affine representations, Amer. J. Math. **102** (1980) 385-407.

[GH1] K.R. Goodearl and D.E. Handelman, Rank functions and $K_0$ of regular rings, J. Pure and Appl. Algebra **7** (1976) 195-216.

[GH2] K.R. Goodearl and D.E. Handelman, Metric completions of partially ordered abelian groups, Indiana U. Math. J. **29** (1980) 861-895.

[H1] David Handelman, **Positive polynomials and product type actions of compact groups,** Memoirs of the Amer. Math.Soc. **320** (1985) 79 p.+ xi.

[H2] David Handelman, Deciding eventual positivity of polynomials, Ergodic Theory and Dynamical Systems **6** (1986) 57-79.

[H3] David Handelman, Integral body-building in $\mathbf{R}^3$, J. of Geometry **27** (1986) 140-152.

[H4] David Handelman, Representing polynomials by linear functions on polytopes, Pacific J. Math. (to appear).

[H5] David Handelman, Stable positivity of orthogonal functions in one variable, preprint (1985).

[HR] David Handelman and Wulf Rossmann, Product type actions of finite and compact groups, Indiana U. Math. J. **33** (1984) 479-509.

[Ho] M. Hochster, Rings of invariants of tori, Cohen-Macauley rings generated by monomials, and polytopes, Ann. Math. 318-337.

[INN] I. Iscoe, P. Ney, and E. Nummelin, Large deviations of uniformly recurrent Markov additive processes, Adv. App. Math. **6** (1985) 373-412.

[K] I. Kaplansky, **Commutative Rings**, Allyn & Bacon (1970).

[McM] P. McMullen, Representations of polytopes and polyhedral sets, Geometriae Dedicata **2** (1973) 83-99.

[N] P. Ney, Convexity and large deviations, Ann. of Prob. **12** (1984) 903-906.

[Ro] R.T. Rockafeller, **Convex Analysis**, Princeton U. Press (1970) Princeton, N.J.

[Sw1] R. Swan, Topological examples of projective modules, Trans. Amer. Math. Soc. **230** (1977) 201-234.

[Sw2] R. Swan, **Algebraic K-Theory**, Lecture Notes in Mathematics **76** (1968) Springer-Verlag.

[ZS] I. Zariski and P. Samuel, **Commutative Algebra** Vol. 1, (1958) D. van Nostrand Co., Princeton.

Mathematics Department
University of Ottawa
Ottawa, Ontario K1N 6N5
Canada

# INDEX

# SYMBOL LIST

## LEMMAS, PROPOSITIONS, THEOREMS, EXAMPLES, etc.